AP* BIOLOGY

. . . .

FLASH REVIEW

OTHER TITLES OF INTEREST FROM
LEARNINGEXPRESS

AP* U.S. History Flash Review

ACT* Flash Review

AP* BIOLOGY
FLASH REVIEW

LEARNINGEXPRESS®

NEW YORK

Printed in the United States of America

9 8 7 6 5 4 3 2 1

First Edition

ISBN 978-1-57685-921-6

For more information or to place an order, contact
LearningExpress at:
    2 Rector Street
    26th Floor
    New York, NY 10006

Or visit us at:
    www.learningexpressllc.com

*AP is a registered trademark of the College Board, which was not
involved in the production of, and does not endorse, this product.

CONTENTS

AP* BIOLOGY

FLASH REVIEW

INTRODUCTION

About the AP Biology Exam

The AP Biology exam tests students' knowledge of core themes, topics, and concepts covered in a typical high school AP Biology course, which offers students the opportunity to engage in college-level biology study. Students who are successful on the AP Biology exam have the opportunity to earn college credit and advanced course placement at thousands of colleges across the country—and success on the exam really helps make a great impression on college admissions committees.

The AP Biology exam is three hours long and consists of the following sections:

- A 100-question multiple-choice section, which students are given 80 minutes to complete. Questions are designed to assess students' knowledge of material covered across the entire AP Biology course (aligned with the course overview provided in this section). The multiple-choice section accounts for 60% of a student's overall exam grade.

- A 10-minute reading period, in preparation for the free-response section of the exam.
- A free-response section, which includes four separate biology questions that students must complete within the 90 minutes allocated. The essays can be based on any of the main content areas covered in an AP Biology course (aligned with the course overview provided in this section), and will assess one's ability to analyze and interpret laboratory and course lecture data and information. The free-response section accounts for the remaining 40% of a student's overall exam grade.

How the AP Biology Exam Is Scored

As previously indicated, the multiple-choice section of the AP Biology exam accounts for

60% of a student's overall score; the remaining 40% is based on performance on the free-response section. These scores are then combined, and the weighted raw scores are turned into a final composite score, which is based on a five-point scale:

5—Extremely well qualified
4—Well qualified
3—Qualified
2—Possibly qualified
1—No recommendation

Typically, a score of three or above indicates that a student has achieved a good grasp of the material covered in his or her AP Biology course.

AP Biology Course Overview

AP Biology courses are designed to cover a variety of specific topic areas and themes. The AP Biology exam is designed to determine how well you have grasped and absorbed the material covered throughout the course. The following is a brief AP Biology course overview.

AP Biology courses—and the AP Biology exam—typically cover the following major topic areas:

Molecules and Cells
Chemistry of Life
- Water
- Organic molecules in organisms
- Free energy changes
- Enzymes

Cells
- Prokaryotic and eukaryotic cells
- Membranes
- Subcellular organization
- Cell cycle and its regulation

Cellular Energetics
- Coupled reactions
- Fermentation and cellular respiration
- Photosynthesis

Heredity and Evolution
Heredity
- Meiosis and gametogenesis
- Eukaryotic chromosomes
- Inheritance patterns

Molecular Genetics
- RNA and DNA structure and function
- Gene regulation
- Mutation
- Viral structure and replication
- Nucleic acid technology and applications

Evolutionary Biology
- Early evolution of life
- Evidence for evolution
- Mechanisms of evolution

Organisms and Populations
Diversity of Organisms
- Evolutionary patterns
- Survey of the diversity of life
- Phylogenetic classification
- Evolutionary relationships

Structure and Function of Plants and Animals
- Reproduction, growth, and development
- Structural, physiological, and behavioral adaptations
- Response to the environment

Ecology
- Population dynamics
- Communities and ecosystems
- Global issues

Across the material covered in a typical AP Biology course, the following major themes are addressed:
- Science as a process
- Evolution
- Energy transfer
- Continuity and change
- Relationship of structure to function
- Regulation
- Interdependence in nature
- Science, technology, and society

In addition to regular course lectures, AP Biology courses include a significant proportion of laboratory work (typically 25% to 30%), which provides students the opportunity to see biology concepts in action and apply what they've learned in hands-on activities.

Detailed information on both the AP Biology exam and the AP Biology course is available on the official College Board website: http://apcentral.collegeboard.com/apc/public/repository/ap-biology-course-description.pdf.

How to Use This Book

AP Biology Flash Review is designed to help you prepare for and succeed on the AP Biology exam. It contains 600 of the most commonly covered terms on the exam, along with their definitions, for quick and effective study and review. On one side of the page are three essential AP Biology terms; on the reverse side are their definitions and/or explanations. The terms are alphabetized for easy access.

Please note that this book is not designed to give an exhaustive review of the entire scope of the terms and concepts inside; it is meant to give you a general overview of biology terms and concepts that you will likely need to know in order to succeed on the AP Biology exam. Being fully aware of the essential terms covered in this book will put you in a great position for success on test day.

Using This Book to Prepare for the AP Biology Exam

AP Biology Flash Review works well as a stand-alone study tool, but it is recommended that it be used to supplement your other course materials and lecture notes as you prepare for the AP exam. The following are some suggestions for making the most of this effective resource as you structure your study plan:

- Do not try to learn or memorize the 600 terms covered in this book all at once. Cramming is not the most effective approach to test preparation. The best approach is to build a realistic study schedule that lets you review around 10 to 15 terms each day. Review a set of terms and then quiz yourself to see how well you've learned them.
- Mark the terms that you have trouble with, so that they will be easy to return to later for further study.
- As you study your course materials in preparation for the AP Biology exam, you may come across terms that are unfamiliar or difficult for you. We recommend that you consult this book

for a quick review of the terms and concepts that often appear on the AP Biology exam.

- Make the most of this book's portability—take it with you for studying on trips, between classes, while commuting, or whenever you have some free time.

ABIOTIC FACTOR

. . . .

ABSORPTION SPECTRUM

. . . .

ACIDIC

Core, inorganic environmental components, including physical, chemical, and geological factors, that have a measurable impact on their respective ecosystems. Abiotic factors have varying effects on living organisms and nonliving elements of the biosphere.

. . . .

The range of discernible electromagnetic radiation, across the electromagnetic spectrum, that a specific substance or absorbing medium allows to pass through. The amount of radiation that passes through is based on the unique chemical makeup of the substance, and serves as an identifying factor.

. . . .

An identifying factor that marks a certain substance as possessing the chemical properties of an acid. An acid is commonly recognized as a substance with a pH lower than 7 that satisfies the Arrhenius, Brønsted-Lowry, and Lewis definitions regarding the chemical properties of an acid:

- Arrhenius acid: a substance that, when dissolved in water, increases the level of H_3O^+ (hydronium ions)
- Brønsted-Lowry acid: a substance that serves as a proton donor during a chemical reaction
- Lewis acid: a substance that accepts an electron pair during a chemical reaction

ACOELOMATES

. . . .

ACROSOME

. . . .

ACTINOPOD

A

Organisms lacking a coelom, the fluid-filled cavity derived from the tissue layers of the body. All known vertebrates and higher animal groups are coelomates; acoelomate organisms include worms and simple animals that have no discernible body cavity.

. . . .

A key component of the male sperm cell; this caplike component covers the anterior head of the sperm cell and contains essential enzymes (including acrosin and hyaluronidase) that serve to penetrate the female ovum, thereby facilitating successful fertilization.

. . . .

A class of known amoeboids (irregularly shaped single-cell organisms) that contain pseudopods (cellular projections made of cytoplasm that are used to facilitate movement and gather food) and are supported by internal cytoskeletons made up of an array of microtubules.

ACTION POTENTIAL

. . . .

ACTION SPECTRUM

. . . .

ACTIVATION ENERGY

A key component of a neuron-firing event in plant and animal cells, which facilitates cellular communication and activates various intracellular processes. Within the plasma membrane of a cell, voltage-gated ion channels are closed while at their resting potential. Once pores in the membrane open, allowing an influx of positively charged ions, these channels will begin to open as the threshold value is reached, thereby increasing the electrochemical gradient. This process is repeated until an explosive action potential is reached, sending a powerful electrical charge that travels through the nerve. A refractory period follows an action potential.

. . . .

Typically, an indication of the wavelength of light that can be used to facilitate a specific chemical reaction. It often refers to the wavelength of light that a plant can use most efficiently to facilitate photosynthesis.

. . . .

The minimum amount of energy needed for a specific chemical reaction to take place. Activation energy (E_a) is measured in kilojoules per mole.

ACTIVE SITE

· ·

ACTIVE TRANSPORT

· ·

ADAPTATION

A

The part of an enzyme (molecules that induce chemical reactions) where the catalytic activity of substrates occurs; these are typically found within lined pockets in the enzyme that also aid in the recognition of binding substrates and participate in chemical reactions.

· ·

Refers to the movement of a chemical substance across and against its concentration gradient (from areas of low concentration to areas of high concentration).

- Primary active transport occurs when the process of active transport requires the use of an energy source, commonly adenosine triphosphate (ATP).
- Secondary active transport uses the electrochemical gradient (where ions pumped out of the cell lead to an electrochemical difference across a membrane) to power the transport process.

· ·

The evolutionary process of acquiring and developing various traits that benefit a specific organism within its environment, including its survival and reproductive functions, by means of natural selection and adaptive plasticity over time.

ADENOID

. . . .

ADHESION

. . . .

ADIPOSE TISSUE

Mass of lymphoid tissue located in the back of the nasal airway, between the nose and the throat; also referred to as the pharyngeal or nasopharyngeal tonsil. This antibody-producing mass protects against infectious agents and potentially dangerous inhaled substances during the first few years of life, after which it typically shrinks and is commonly removed during tonsillectomies.

. . . .

The intermolecular forces that lead to the binding of disparate particles or molecules. This can occur through the processes of chemical adhesion, diffusive adhesion, dispersive adhesion, electrostatic adhesion, or mechanical adhesion.

. . . .

Fat-storing connective tissue that serves to store energy (lipids), as well as to cushion and insulate the body. Adipose tissue can be found subcutaneously and surrounding internal organs. In animals it exists in two forms—white adipose tissue (WAT) and brown adipose tissue (BAT).

AEROBE

. . . .

AGAR

. . . .

ALDEHYDE

A

An organism that requires an oxygen source for survival and growth.

. . . .

A gelatinous nutrient source for various types of growing microorganisms, often used to sustain specimens and cultures during lab work. It is typically derived from the boiling of marine algae, and has also served a variety of practical uses throughout history.

. . . .

A common organic compound that contains a formyl group and has a carbon skeleton with a carbonyl center. Aldehydes are often used in manufacturing processes, including the creation of fragrances and dyes.

ALKALINE

. . . .

ALLANTOIS

. . . .

ALLELE

A

A substance with a pH greater than 7 that neutralizes acids; the effectiveness with which an alkaline neutralizes an acid is referred to as its alkalinity.

· ·

An identifying embryonic component of all amniotes, this saclike membrane is designed to process liquid waste and facilitate gas exchange in the developing embryo.

· ·

One form of a gene located on a chromosome. Alleles can be dominant or recessive, and variations in alleles can result in variations in phenotype, or observable traits.

ALLERGEN

. . . .

ALLOMETRY

. . . .

ALLOPATRIC SPECIATION

A substance (antigen) that can trigger an allergic response in an organism. There is a broad spectrum of allergens (including food sources, pollen, dust mites, pet dander, and fungal allergens) that can lead to a range of hypersensitive responses, depending on organism sensitivity. Common treatments for allergen sensitivity include immunotherapy, antihistamines, and nasal decongestants.

. . . .

The study of an organism's body size in relation to its various parts. Research in this field has yielded significant insight into intra-organism growth rates and dimension ratios, as well as statistical shape analysis, shape variation among individuals, and comparisons across species.

. . . .

The process of biological species formation that occurs when same-species populations are isolated, often due to geographic or social barriers or changes. Over time, these isolated populations may develop observable and distinctly different traits and characteristics. These variations are typically due to selective or adaptive changes, allelic drift, and/or mutations in the gene pool. This may lead to populations that are too distinct to mate, even if the original barriers or changes are removed, which can eventually lead to the emergence of completely new species.

ALLOSTERIC REGULATION

· ·

ALPHA CELLS

· ·

ALVEOLUS

A

The process of enzyme regulation that occurs when an effector molecule binds at a specific site (allosteric site), thereby altering the enzyme's receptivity to the substrate—either by making it more receptive (allosteric activation) or less receptive (allosteric inhibition) at the active sites.

· ·

Cells located in the islets of Langerhans, where the endocrine cells of the pancreas are located. Alpha cells produce glucagon, a peptide hormone that is responsible for increasing blood glucose levels.

· ·

A small cavity found in the lungs of mammals, made up of an epithelial layer and containing collagen and elastic fibers. Alveoli facilitate the exchange of carbon dioxide and oxygen out of and into the bloodstream through the process of diffusion, which is an essential component of respiration.

AMENSALISM

. . . .

AMINO ACIDS

. . . .

AMINOACYL-tRNA

A

An aspect of population interaction wherein one population or organism adversely affects another, and the first population remains unaffected. This is often due to the biological products of one affecting the other, or can possibly be the result of overt behavior.

· ·

Organic compounds, discovered in the early 19th century, found in animal and plant tissues that are among the core building blocks of proteins. Amino acids consist of both carboxyl (COOH) and amino (NH_2) groups, and the human body can synthesize both essential and nonessential amino acids. They also have a variety of industrial applications, including drug and plastic manufacturing.

· ·

A transfer ribonucleic acid, which facilitates the attachment of an amino acid to a ribosome for incorporation into a polypeptide chain that is being created.

AMMONIFICATION

....

AMNION

....

AMNIOTE

A

An essential component of the decomposition process, wherein organic matter or waste, existing as nitrogen, is broken down into ammonium (NH_4^+).

· ·

A protective membrane surrounding mammalian, reptile, and bird embryos, which helps facilitate their growth and development.

· ·

A classification of vertebrate animals that develop amniotic eggs, which have embryos surrounded by protective membranes such as amnions. The first known terrestrial amniotes appeared on Earth approximately 340 million years ago.

ANABOLISM

. . . .

ANAEROBE

. . . .

ANAEROBIC

A

A metabolic phase at the cellular level, in which small and simple units of matter that are the building blocks of complex living tissue are synthesized into larger and more complex molecules.

. . . .

An organism that does not require oxygen for development or survival. Anaerobes are separated into three types:

1. Aerotolerant organisms are fermentative and do not require oxygen, but can survive and utilize aerobic respiration if oxygen is present.
2. Facultative anaerobes do not require oxygen, but can use it if it is in the environment.
3. Obligate anaerobes utilize anaerobic respiration or fermentation and are adversely affected by the presence of oxygen.

. . . .

Directly translated as "without oxygen," anaerobic organisms do not require, or should not be near, an oxygen source for survival and growth.

ANAPHASE

. . . .

ANDROGEN

. . . .

ANEUPLOIDY

A

The third stage of cell division (both meiosis and mitosis). During anaphase, the cell's spindle fibers stretch the cell into an oval shape, and the chromosomes split and migrate to opposite ends of the cell.

. . . .

A natural or synthetic male steroidal hormone (i.e., androsterone, androstenedione, dihydrotestosterone, and testosterone) that regulates the growth, development, and maintenance of an organism's masculine traits, characteristics, and reproductive system.

. . . .

A type of abnormality that occurs during defective chromosome separation in cell division, resulting in an abnormal number of chromosomes (either a missing or an extra chromosome on a set). This is a common cause of genetic disorders and birth defects.

ANGIOSPERM

. . . .

ANNELID

. . . .

ANTIBODY

A

The largest and most diverse plant group; refers to any seed-producing, flowering plant with functioning ovules contained in its ovaries. Derived characteristics of angiosperms include flowers, stamens with pollen-sac pairs, and endosperm.

· ·

Worms and worm-like organisms contained in the phylum Annelida, including earthworms and leeches, that are characterized by ringed, segmented bodies in an elongated cylindrical shape.

· ·

A Y-shaped immunoglobulin, produced on the surface of B cells, that is secreted into the bloodstream and used by organisms during immune system responses, including identifying, tagging, and neutralizing target foreign antigens (e.g., potentially harmful viruses and bacteria).

ANTICODON

. . . .

ANTIGEN

. . . .

ANTIOXIDANT

A

A triplet nucleotide sequence on a tRNA molecule; its purpose is to bind with a corresponding and complementary mRNA molecule codon during protein synthesis.

· ·

A substance that invades a host organism and triggers an immune response (i.e., the production of antibodies). Antibodies bind to respective antigens at designated antigen binding sites in an effort to neutralize or destroy the foreign invader. Antigens can be exogenous (entering the body from the outside) or endogenous (generated from within an organism's cells).

· ·

A chemical compound that serves to inhibit the process of oxidation (the transfer of hydrogen or the loss of electrons to an oxidizing agent), which can prove harmful to an organism if free radicals are produced as a result; free radicals can lead to cellular chain reactions that damage or kill healthy cells.

APICAL DOMINANCE

. . . .

APOMIXIS

. . . .

APOPLAST

A

An occurrence in plants wherein the terminal bud of a plant's shoot (a plant's main central stem) exhibits extreme growth dominance and inhibits the development of lateral buds. This often occurs as the result of an apical bud producing and releasing the hormone auxin, which inhibits lateral bud growth.

. . . .

A term often used in botany to describe asexual reproduction in plants (reproduction without fertilization). Offspring produced through apomixis are genetically identical to their parents.

. . . .

The structural component of a plant that exists outside of the plasma membrane. This includes the intracellular spaces and materials, and cellular walls. Water and solutes flow unopposed through the plant's apoplastic route. Carbon dioxide (CO_2) is solubilized in the apoplast prior to photosynthesis.

APOPTOSIS

. . . .

ARBOREAL

. . . .

ARCHAEA

A

The process of cellular death and disintegration, programmed through either intracellular or extracellular cell signals, which serves as a natural component of growth and development in multicellular organisms. In an average human, tens of billions of cells undergo apoptosis daily.

. . . .

Refers to tree-dwelling biological organisms, or animals that frequent trees (e.g., arboreal monkeys).

. . . .

A domain of simple, single-celled microorganisms, structurally similar to bacteria, that are characterized by their lack of a cell nucleus and any discernible cellular organelles bound by membranes.

ARTERIOLE

. . . .

ARTERIOSCLEROSIS

. . . .

ARTHROPOD

A

A blood vessel within an organism's vasculature that is characterized by its small diameter and smooth muscular walls. Arterioles are key components of a body's microcirculation and branch out from arteries to capillaries; vascular resistance occurs within the arterioles.

· ·

A cardiovascular disease resulting from the hardening of the arteries and the loss of essential arterial elasticity within an organism. This can lead to a potentially dangerous restriction of blood flow within the body, denying essential organs vital oxygen and nutrients.

· ·

An invertebrate animal (including arachnids, crustaceans, and insects) that is characterized by segmented body parts and jointed limbs within an exoskeleton. Over a million observable species fall into this classification, and they vary widely in size, shape, and structure.

ARTIFICIAL SELECTION

. . . .

ASCUS

. . . .

ASSIMILATION

A

The process of selective breeding in biological organisms in an effort to intentionally bring about the development and proliferation of desired traits (as opposed to natural selection, in which the non-random interplay between organisms and their environments determines the biological traits that benefit survival, without the intervention of humans). Examples of artificial selection can range from animal and plant breeding to genetic engineering.

• •

A key characteristic of ascomycete fungi; the sac-like, sexual spore-bearing cell exists in four basic types:
1. Unitunicate-operculate asci have spores that release from lid-like operculums.
2. Unitunicate-inoperculate asci have spores that release from an elastic ring that acts like a pressure valve.
3. Bitunicate asci have double walls that facilitate the release of ripened spores.
4. Protounicate asci are spherical and their walls dissolve when ripened spores are ready to be released.

• •

A biological process that breaks down, converts, and delivers vital nutrients in usable forms to cells, tissues, and organs. Some common types of biological assimilation include photosynthesis, nitrogen fixation, and nutrient absorption from food following digestion.

ASSOCIATIVE LEARNING

. . . .

ASSORTATIVE MATING

. . . .

ATHEROSCLEROSIS

A

A learned association involving two stimuli or a stimulus and a behavior. Associative learning exists in two forms:
1. Classical conditioning is a learned association between a stimulus and a reflex-inducing behavior, often through repeated exposure.
2. Operant conditioning is a learned associative behavior as a result of positive consequences (reinforcement) or negative consequences (punishment).

· ·

Non-random mating behavior in which individuals mate with those who possess certain similar traits or characteristics more frequently than what would occur during random mating.

· ·

A potentially health-threatening condition that occurs when the walls of the arterial blood vessels in an organism thicken, harden, and narrow due to the accumulation of cholesterol and other fats, leading to plaque buildup. Over time, this can lead to cardiovascular disease.

ATP (ADENOSINE TRIPOSPHATE)

. . . .

AUTOSOME

. . . .

AUTOTROPH

A

A nucleoside triphosphate compound that serves as the chief energy-transferring coenzyme for cells. A variety of cellular processes are fueled by ATP, including active transport, cell division, metabolism, locomotion, and synthesis.

. . . .

A type of non-allosome chromosome that is not involved in sex determination or the development of sexual characteristics. The average human possesses 22 autosome pairs (along with one allosome pair).

. . . .

A type of organism that obtains its energy (organic compound creation, including proteins, fats, and carbohydrates) from simple substances in its immediate surroundings, not by eating other organisms. Types of autotrophs include:

- Phototrophs: Acquire energy from sunlight through photon capture.
- Lithotrophs: Acquire energy from inorganic compounds.
- Chemotrophs: Acquire energy from electron donor sources.

AUXIN

. . . .

AV NODE

. . . .

AXILLARY BUD

A

A type of plant hormone that plays crucial roles in a variety of plant growth, development, and behavioral processes. These include leaf, flower, and fruit development, root and cell growth, and hormonal activity regulation. It also plays an active role in hydrotropism, geotropism, and phototropism.

. . . .

The atrioventricular (AV) node is a grouping of specialized tissue fibers that conduct electrical impulses between the atrial and ventricular chambers of the heart, helping to coordinate proper heart functioning and normal cardiac rhythm. Blood supply to the AV node comes from the posterior interventricular artery.

. . . .

A part of a plant, it is an embryonic shoot that is located between the plant's stem and leaf blade. Axillary buds can develop into flowers under the right conditions, and can also help in plant identification (single-leafed vs. multi-leafed).

AXON

. . . .

A

A structure within a neuron that serves to
conduct electrical impulses from the soma (cell
body) to target cells. These long, thin proto-
plasmic projections are the nervous system's
chief transmission tools, transmitting essential
signals to cells within an organism across synaptic junctions. Axons are typically covered in
a myelin sheath to facilitate efficient and rapid
electrical impulse transmission.

• •

BACTERIOPHAGE

. . . .

BALANCED POLYMORPHISM

. . . .

BASE

A type of virus that is extremely prolific in the biosphere and invades bacterial cells; phages are sometimes used in place of antibiotics and to combat drug-resistant bacterial strains.

. . . .

A type of evolutionary selection designed to preserve elevated frequencies of multiple beneficial alleles within a gene pool over time, in order to benefit the continued development of a population.

. . . .

Substance with a pH level greater than 7 that is a proton acceptor (can accept hydrogen ions and donate electron pairs), and that can serve as a catalyst in a variety of chemical reactions. Neutralization occurs when an acid and a base interact.

BASEMENT MEMBRANE

. . . .

BASIDIUM

. . . .

BATESIAN MIMICRY

B

A structure of the epithelial membrane consisting of a thin set of fibers in two layers (the basal lamina and the reticular lamina). The basement membrane lines various organ surfaces and cavities, anchors the epithelium to surrounding connective tissue, and serves as a barrier from malignant cells.

. . . .

An appendage found on the gill margins of basidiomycota fungi that produces spores (basidiospores) for sexual reproduction. Spores are typically released from the sterigma (stalk at the top of the basidium) upon maturity.

. . . .

A type of evolutionary adaptation in which a relatively harmless species acquires traits that allow members to imitate the characteristic warning signals of a more dangerous species in an effort to protect themselves from predators. Batesian mimicry can utilize each sense or any combination of the senses.

BETA CELL

· ·

BILATERAL SYMMETRY

· ·

BINARY FISSION

A cell located in the pancreas that is designed to store and release insulin (to regulate blood glucose levels), C-peptides (a protein that helps regulate healthy organ function), and amylin (contributes to glycemic regulation).

. . . .

A type of biological symmetry wherein a central longitudinal plane (sagittal plane) connects and divides an organism into two symmetrical halves; however, the symmetry is often a rough approximation and not a perfect mirror image.

. . . .

A type of cellular subdivision that divides a cell into two separate cells. It is a process of asexual reproduction that involves the replication and separation of the original genetic material into an equal and distinct copy.

BIOENERGETICS

....

BIOMASS

....

BIOME

B

A discipline that focuses on the study of how organisms regulate their energy production and sources, particularly involving the role of energy in biological processes, biological energy synthesis and transformation, and the creation of molecular chemical bonds.

· ·

Biomass can refer to either the actual mass of a particular subset of organisms in a predetermined ecosystem, or organic matter that can be used as a source of energy, often through biochemical, chemical, or thermal conversion processes.

· ·

A major global ecological community that often occupies a geographically distinct region with organisms that adapt to that respective environment; often divided into terrestrial biomes (i.e., desert, grassland, tundra, tropical rainforests, etc.) and aquatic biomes (freshwater and marine).

BIOSPHERE

. . . .

BIVALVE

. . . .

BLASTOCOEL

A broad term encapsulating the sum of the planet's ecosystems, including all living things and their respective interrelationships. Earth's biosphere is divided into a number of distinct, observable biomes.

. . . .

A classification of marine and freshwater mollusk that consists of a hinged shell with two distinct valves that protect a soft inner body. Bivalves vary greatly in size and shape, and include clams, mussels, oysters, and scallops.

. . . .

An embryonic structure that forms in the early stages of embryo growth; it is the fluid-filled central cavity of the blastocyst.

BLASTOCYST

. . . .

BLASTULA

. . . .

BLOOD-BRAIN BARRIER

A mammalian structure that develops early in the formation and development of an embryo; it contains the group of inner cells referred to as the embryoblast and an outer cell layer covering the embryoblast and blastocoel called the trophoblast.

· ·

A cellular sphere that develops during animal embryo formation (during the end of the cleavage process, when the zygote undergoes cell division).

· ·

A physiological construct of dense cells designed to regulate the permeability of brain capillaries (separating blood from extracellular fluid), thereby protecting brain tissue and facilitating the healthy functioning of the central nervous system. The barrier restricts the diffusion of microscopic material and also transports beneficial metabolic products.

BRAIN STEM

. . . .

BRONCHIOLE

. . . .

BRYOPHYTE

The small, stalk-like posterior portion of the brain in vertebrates that adjoins the spinal cord and includes the medulla oblongata, pons, and midbrain. The brain stem serves to regulate the central nervous system, is a key component of motor and sensory system nerve connections between the brain and body, and is involved in normal cardiac and respiratory functioning.

. . . .

One of the small respiratory tubes where air passes through from the mouth and nose to the lungs' alveoli. The widths of the bronchioles fluctuate (bronchodilation or bronchoconstriction) to regulate air flow.

. . . .

A classification of non-vascular land plant (embryophyte) that lacks xylem and phloem for internal water circulation. Bryophytes can be either dioicous or monoicous, and include mosses, hornworts, and liverworts.

BUDDING

....

BUFFER SOLUTION

....

A type of asexual reproduction in which an outgrowth from an organism develops and forms a new, genetically identical organism, detaching from the parent organism upon maturity. Budding occurs in simple unicellar organisms (yeasts, etc.) as well as more complex organisms (hydras, corals, sponges, etc.).

. . . .

A solution containing a conjugate acid-base pair that serves to keep pH levels stable in various chemical reactions within organisms, as well as in industrial processes.

. . . .

CALORIE

. . . .

CAPSID

. . . .

CARBON FIXATION

C

A unit of energy (typically food energy). Small calorie is defined as the amount of energy that is required to increase the temperature of 1 gram of water by 1°C; large calorie, or kilocalorie, is defined as the amount of energy that is required to increase the temperature of 1 kilogram of water by 1°C.

. . . .

The encapsulating protein shell of a virus. Capsids can vary in structure, though most are helical or icosahedron in shape. They protect the genetic material of the virus.

. . . .

The process by which autotrophs (plants, etc.) reduce carbon (typically CO_2) into a usable organic compound; photosynthesis is the most common method of carbon fixation.

CARBONYL GROUP

. . . .

CARBOXYL GROUP

. . . .

CARCINOGEN

C

A functional group consisting of a carbon atom and an oxygen atom that are joined by a double bond (C=O); a wide variety of compounds contain a carbonyl group, including aldehyde, amide, carboxylic acid, ketone, and ester.

. . . .

A functional group consisting of a double bond between a carbon atom and an oxygen atom (C=O), bonded to a hydroxyl group (OH). Carboxylic acids containing at least one carboxyl group include acetic acid (CH_3CO_2H), benzoic acid ($C_7H_6O_2$), chloroacetic acid ($ClCH_2CO_2H$), formic acid (HCO_2H), and oxalic acid ($H_2C_2O_4$).

. . . .

A substance or agent that produces cancer as a result of altering the genetic structure of cells. Carcinogenic substances can be either natural or synthetic; known carcinogens include asbestos, DDT, and tobacco smoke.

CARPEL

. . . .

CARRYING CAPACITY

. . . .

CATABOLISM

A flowering plant's female reproductive organ, which includes the ovary, stigma, and style. They are the components of the gynoecium, all of the carpels within a flower, which can consist of single or multiple carpels.

· ·

The maximum size of a population within a biological species that a specific environment can successfully maintain, given the amount of available vital resources (i.e., food, water, habitat, etc.). A population increase beyond carrying capacity can lead to adverse conditions and a negative impact on both individual organisms and the group as a whole.

· ·

The metabolic processes within an organism that serve to break down large, complex molecules into simpler parts, thus releasing energy. Some examples of catabolism include:
- polysaccharides → monosaccharides
- lipids → fatty acids
- nucleic acids → nucleotides
- proteins → amino acids

CATALYSIS

. . . .

CELLULAR DIFFERENTIATION

. . . .

CELLULOSE

The change in a chemical reaction's rate as a result of the introduction of a catalyst. Catalysts can either increase the speed of a chemical reaction (positive catalyst) or slow down the rate of a chemical reaction (inhibitor). Furthermore, there is a classification of substances that directly impact the effectiveness of catalysts, either increasing their activity (promoters) or decreasing or deactivating catalysts (catalytic poisons).

. . . .

A phase of development within a multicellular organism where cells change, often due to gene expression, and become more structurally and functionally specialized, thereby altering their membrane potential, metabolic activity, size, shape, and signal responsiveness.

. . . .

A polysaccharide and the most abundant organic compound on Earth. It has a variety of organic uses (i.e., it serves as the main component of green plant cell walls and can be digested by some animals, etc.), as well as industrial applications (i.e., paper and textile production, etc.).

CENTRAL NERVOUS SYSTEM

. . . .

CENTRIOLE

. . . .

CENTROMERE

C

The anatomical system within the dorsal cavity of bilateral multicellular animals that includes the brain and the spinal cord. The central nervous system (CNS) coordinates and regulates information and resulting behaviors throughout the body.

. . . .

A cylindrical component of most animal cells, composed of microtubules, that helps facilitate cytokinesis during mitosis. Centrioles also play a crucial role in cellular arrangement and organization.

. . . .

The centralized structure within a chromosome that serves to link pairs of sister chromatids. During cell division, it plays a key role in the structuring of the kinetochore, attaching to the developing spindle fibers prior to anaphase.

CENTROSOME

. . . .

CEPHALIZATION

. . . .

CEPHALOPOD

A structural component of animal cells, composed of two centrioles and a protein mass known as the pericentriolar material. The centrosome is the primary organizing and formation hub of a cell's microtubules, and serves to regulate the normal progression of the stages of cell division.

• •

The evolutionary development and centralization of nervous tissue and sensory organs in the anterior (head) region of organisms.

• •

A diverse grouping of marine animals that are members of the class Cephalopoda. These highly developed and social invertebrates are often identifiable by their large heads, excellent vision, tentacles, and bilateral symmetry. Cephalopods cannot survive in freshwater environments.

CHAPARRAL

. . . .

CHARACTER DISPLACEMENT

. . . .

CHELICERAE

A scrubland plant environment characterized by the dense growth of drought-resistant plant life and spiny shrubbery with small, hard leaves; chaparral thrives in dry, hot summers and mild, wet winters.

· ·

A population phenomenon that occurs when similar species that live in close proximity have their differences accentuated, but these differences are then minimized when those same species do not live in close proximity. This is aligned with the competitive exclusion principle and the hunt for limited resources.

· ·

The pointed mouth appendages of organisms in the arthropod subphylum Chelicerata (i.e., arachnids, etc.) used for grasping or injecting venom into prey.

CHEMIOSMOSIS

. . . .

CHEMOAUTOTROPH

. . . .

CHEMOHETEROTROPH

C

An ATP-generating process in which hydrogen ions move across a semipermeable membrane and down their electrochemical gradient, from a high to a low proton density.

• •

An organism (commonly bacteria) that can utilize carbon dioxide (CO_2), as well as inorganic energy sources, to synthesize the organic compounds it requires. Scientists theorize that Earth's complex organisms evolved from simple chemoautotrophs.

• •

An organism that cannot use carbon fixation to create necessary organic compounds and energy; it must rely on the consumption of organic or inorganic molecules.

CHEMORECEPTOR

. . . .

CHEMOTROPH

. . . .

CHIASMA

C

A type of sensory receptor that transmits key information via chemical signal (following an action potential) in response to various environmental stimuli that affect the sensory organs; chemoreceptors are also involved in the regulation of heart rates and breathing rates.

. . . .

A type of organism that creates the energy it needs for functioning through oxidation (loss of electrons during a redox reaction) of organic or inorganic molecules. Chemo-organotrophs utilize organic molecules, and chemolitho-trophs utilize inorganic molecules.

. . . .

The point where two chromatids cross and exchange genetic material during meiosis. Chiasmata between two homologous non-sister chromatids results in genetic changes (known as "crossing over"), whereas chiasmata between two sister chromatids do not cause changes because they are genetically idential.

CHITIN

....

CHLOROPHYLL

....

CHLOROPLAST

A common glucose derivative that is found throughout nature, including the exoskeletons of various types of arthropods (i.e., crustaceans, insects, etc.) and fungi cell walls. Chitin also has a variety of medical, industrial, and agricultural applications.

• •

A type of pigment (colored green) found in plants, it is critical in the synthesis of solar energy to chemical energy, also known as photosynthesis. Chlorophyll is found within the chloroplasts of plants, where photosynthesis takes place, and utilizes the process of resonance energy transfer to transport absorbed light to a reaction center chlorophyll pair within a plant's photosystem, where the process of energy synthesis occurs.

• •

A cellular structure (a type of plastid) within plants where the process of photosynthesis takes place. Light (largely solar energy) is captured and absorbed in thylakoids of chloroplasts, where it is converted into a usable energy source through the process of resonance energy transfer.

CHOANOFLAGELLATE

. . . .

CHORDATE

. . . .

CHORION

C

A spherical unicellular structure, consisting of an ovoid cell body and flagellum encased in microvilli, whose process of obtaining usable nutrients from the environment plays a critical role in the maintenance of the biosphere's carbon cycle. They are also viewed as a critical evolutionary link between animals and unicellular organisms, and are important for research and understanding of how complex organisms developed over time.

. . . .

A classification of animals belonging to the Chordata phylum; they are characterized by an elongated dorsal nerve cord (known as a notochord) that appears in the embryonic stage, a dorsal neural tube (that develops into a spinal cord), filtering pharyngeal slits, and a postanal tail.

. . . .

An essential extraembryonic membrane layer in mammals that is part of the developing placenta; it consists of two layers—an inner cell layer (mesoderm) and an outer trophoblast that serves to supply nutrients to the embryo from the mother's blood.

CHROMATID

. . . .

CHROMATIN

. . . .

**CHROMOSOME
NONDISJUNCTION**

A strand (one of two) of a replicated chromosome; the two sister chromatids (joined at the centromere) can be either homozygous or heterozygous.

. . . .

Within a eukaryotic cell's nucleus, chromatin is the mix of genetic material—proteins and DNA—that form chromosomes during the process of cell division. During cell division the structure of chromatin changes significantly, to facilitate chromosome development and to help package, strengthen, and protect DNA.

. . . .

A breakdown during the process of cell division (meiosis or mitosis) in which the chromosome pairs fail to properly separate, resulting in aneuploidy—an abnormal number of chromosomes that often leads to genetic disorders.

CHYME

....

CHYTRID

....

CILIA

The liquefied form of ingested food after the stomach crushes and mixes it during the process of digestion.

• •

One of the primitive fungi (usually aquatic) that travel via the use of protruding flagella.

• •

Lashlike components (made up of microtubules) in eukaryotic cells that exist in motile or non-motile (primary) forms; motile cilia serve to facilitate locomotion and transport, and primary cilia are key components of sensory organs, facilitating cell growth, chemical sensation, and signal transduction.

CILIATE

. . . .

CIRCADIAN RHYTHM

. . . .

CLADISTICS

A protozoan that is identifiable by the existence of cilia, lashlike components (made up of microtubules) in eukaryotic cells used for transportation, feeding, attachment, and gathering sensory information from the environment. Ciliates are common in all types of water formations, from ponds to oceans, and vary in size and structure. They contain a nucleus for reproduction (a small, diploid microneucleus) as well as for regulation (polyploidy macronucleus), feed on bacteria and algae, and can reproduce through asexual or sexual reproduction.

· ·

A biological or physiological process or cycle in eukaryotic organisms that takes approximately 24 hours to complete; although these rhythms are endogenous and do not require external cues to persist, they can be entrained (affected by external cues such as daylight for the sleep cycle).

· ·

A classification method that groups organisms into clades (including an ancestor organism along with its descendants) based on associated chronological order of development rather than by the common method of grouping via shared derived characteristics.

CLADOGRAM

. . . .

CLEAVAGE

. . . .

CLONAL SELECTION

C

A diagramming system based on the classification methodology used in cladistics. Cladograms can exist either horizontally or vertically, with branches that designate ancestors and descendant organisms, and show common ancestry.

· ·

The rapid succession of cell division in early embryonic development, during which a zygote is transformed into a blastula (a hollow sphere of clustered cells called blastomeres). The blastula remains approximately the same size as the zygote, despite the increase in number of cells.

· ·

A model regarding how immune systems function in response to infections and antigens. According to clonal selection, lymphocytes have a single receptor type and are selectively triggered and activated by specific antigens.

CLOSED CIRCULATION

. . . .

CLUMPED DISTRIBUTION

. . . .

COCCUS

C

A type of circulation, typically referring to the flow of blood in vertebrates when referred to in biological terms, wherein blood within an organism stays in the circulatory system as it travels and is confined to the vessels of the body.

· ·

A type of population dispersion characterized by a minimal distance between neighboring organisms in relation to the location of essential resources. Clumped distributions typically occur in nature when a scarce supply of resources causes animals to gather around them in close proximity.

· ·

The term for a spherically shaped bacterium; one of three possible shape classifications for bacteria (the other two being baccilus, or rod-shaped, and spirillum, or spiral-shaped).

CODOMINANCE

. . . .

CODON

. . . .

COELOM

C

A relationship between alleles of a gene wherein both are clearly expressed and one is not dominant over the other in the phenotype of a heterozygous organism.

. . . .

The basic structural foundation of an organism's genetic code, consisting of three nucleotides that join to make up the individual RNA and DNA units. During protein synthesis, codons serve to specify which amino acid will be added next, underlying the genetic code.

. . . .

A structure formed within the mesoderm germ layer of certain animals; a fluid-filled cavity in various mammals that forms the peritoneal, pleural, and pericardial cavities.

COENOCYTE

....

COENZYME

....

COFACTOR

A cell that contains more than one nucleus (multinucleate) as a result of multiple nuclear divisions without undergoing the accompanying process of cytokinesis (cytoplasm division forming two separate cells) occurring.

. . . .

Cofactors that are loosely bound to proteins (usually enzymes) during a biological process to help facilitate catalysis of chemical reactions.

. . . .

A substance that binds to a protein (often enzymes), that is either necessary or helpful during catalysis of a chemical reaction. Cofactors can be either organic or inorganic, loosely bound or tightly bound, and sometimes several cofactors are required by enzymes.

COHESION

. . . .

COLEOPTILE

. . . .

COLLENCHYMA

C

The force of attraction that binds like molecules and forms substances, often through the formation of hydrogen bonds. The process of cohesion underlies several foundational physical phenomena, including surface tension and capillary action.

. . . .

A structural component of a germinating monocotyledon (monocot); the protective pointed sheath covering the growing shoot and embryonic leaves.

. . . .

A type of plant tissue made up of thick, uneven walls of elongated cells that serve to provide structural support and protection to young and growing plants. There are three types of collenchyma: angular (thicker at intracellular contact points), lacunar (thicker at points proximal to intracellular spaces), and tangential (thicker at the tangential face of ordered cell wall rows).

COLUMNAR EPITHELIA

. . . .

COMMENSALISM

. . . .

COMPANION CELL

C

A type of epithelial cell that is column-shaped, with a height that is several times longer than its width. They can be unilayered (simple) or multilayered (stratified).

. . . .

A type of symbiotic relationship that exists between two organisms in which one of the organisms benefits from the arrangement and the other neither benefits nor is harmed.

. . . .

A type of parenchyma plant cell that is connected to and facilitates the proper functioning of the sieve-tube element. Companion cells are connected to the sieve-tube element at the cell's cytoplasm (by plasmodesmata).

COMPETITIVE INHIBITOR

. . . .

COMPLEMENT SYSTEM

. . . .

CONCENTRATION GRADIENT

C

A substance that serves to decrease or inhibit enzyme activity by connecting to the enzyme's active site, thereby preventing it from binding with the substrate.

. . . .

A group of beneficial blood proteins that are a part of the immune system and support the functioning of antibodies to fight potentially harmful pathogens. The complement system is regulated by a series of complement control proteins, and is activated by three biochemical pathways—the alternative pathway, the classical pathway, and the lectin pathway.

. . . .

The difference in concentration of a chemical substance or solute; a concentration gradient triggers movement from areas of higher concentration to areas of lower concentration.

CONDENSATION REACTION

. . . .

CONIFER

. . . .

CONJUGATE BASE

C

An intermolecular chemical reaction that includes the bonding of two molecules or functional groups and results in the loss of a small molecule (known as a dehydration reaction when water is lost, but can also lead to the loss of acetic acid, hydrogen chloride, methanol, etc.). A variety of biological synthesis processes are condensation reactions, and they also have a variety of useful industrial and research applications.

• •

A type of cone-bearing, seed-producing plant (gymnosperm) that dates back approximately 300 million years and includes a variety of woody plants and trees, many of which exhibit apical dominance and secrete resin, such as cypresses, firs, pines, and spruces.

• •

A molecule or ion that has the ability to gain or accept a proton (H^+; hydrogen cation) following an acid-base chemical reaction; a key component of Brønsted-Lowry acid-base theory.

CONSANGUINEOUS MATING

....

CONTINENTAL DRIFT

....

CONTINUOUS VARIATION

Refers to the mating of individuals who are closely biologically related by blood. Offspring from consanguineous mating are more likely to exhibit recessive or deleterious traits and have physical and/or health defects.

．．．．．．．．．．．．．．．．．．．．．．．．．．．．

The movement and subsequent global positioning of the major land masses (continents) of the Earth over time, due to drifting across the planet's oceans; the theory of continental drift is supported by a wealth of research, including the location of various plant and animal fossils as well as work done in the area of plate tectonics.

．．．．．．．．．．．．．．．．．．．．．．．．．．．．

Within a population, the gradation of minor differences regarding a particular trait whose expression is genetically dependent (e.g., height).

CONTRACTILE VACUOLE

. . . .

CONVERGENT EVOLUTION

. . . .

COREPRESSOR

A cellular substructure that is instrumental in maintaining the homeostasis of an organism's water level and concentration through osmotic pressure regulation (osmoregulation). Contractile vacuoles are typically found in protists and unicellular algae.

. . . .

The evolutionary acquisition of a similar, environmentally beneficial, analogous trait or body form across diverse, unrelated species. Convergent evolution can occur if species evolve in similar environmental conditions, possess similar ecological niches, and/or have similar ancestral lineages.

. . . .

A substance (small molecules in prokaryotes; proteins in eukaryotes) that serves to inhibit gene expression by triggering the transcription factors of repressors with which they bind.

CORTICOSTEROID

. . . .

COTRANSPORT

. . . .

COUNTERCURRENT FLOW

C

A class of chemical compounds that serves a variety of purposes in organisms, including involvement in behavioral response, blood electrolyte levels, carbohydrate metabolism, and immune stress responses. Corticosteroids are naturally produced in organisms (synthesized in the adrenal cortex of vertebrates) but can also be developed in labs, administered topically or inhaled, and used for treating a variety of conditions. Corticosteroids are typically grouped into four classes—hydrocortisones, acetonides, betamethasones, and esters.

. . . .

The passing of molecules or ions, in a defined ratio, across biological membranes from an area of low concentration to an area of greater concentration, against its concentration gradient (also referred to as secondary active transport or coupled transport).

. . . .

The process by which the flow of two bodies in opposite directions, often set up in a continuing circuit or loop, facilitates the exchange or transfer of some property or item (often heat or energy). The two flowing bodies are often liquids, but can exist as solids or gases. Countercurrent exchange can be found in both biological and industrial systems.

COVALENT BOND

. . . .

CRISTA

. . . .

CROSS-FERTILIZATION

C

A type of chemical bond that occurs when atoms of comparable electronegativity share electron pairs, and a balance of both repulsive and attractive forces is struck.

. . . .

In a mitochondrion, a pocket within the inner membrane that contains the electron transport train as well as proteins, creating optimal conditions for chemical reactions to occur, including cellular respiration and synthesis of ATP for energy (through chemiosmosis).

. . . .

The process of fertilization in organisms that reproduce sexually; the term refers to the fusion of gametes during biological reproduction. In animals, this entails ova being fertilized by spermatozoa; in plants, it entails a flower from one plant being fertilized by the pollen of another plant.

CUBOIDAL EPITHELIA

. . . .

CYANOBACTERIA

. . . .

CYTOKINES

C

A type of densely packed cell that makes up the structure of the epithelium—one of the basic types of animal tissue—that is characterized by its cube-like shape, with an approximately equal height-to-width ratio.

- Simple cuboidal epithelia exist in single layers and are found on ovaries, in nephron lining, within the eye and thyroid, and on the walls of renal tubules.
- Stratified cuboidal epithelia exist in multiple layers and serve to protect sweat gland ducts, mammary glands, and salivary glands.

. . . .

A type of bacteria, named for its blue-green color, that utilizes photosynthesis to obtain energy. Cyanobacteria thrive individually or in colonies, can be found in nearly every type of habitat in the biosphere, and are thought to have played a key role in the development of Earth's biodiversity and proliferation of oxygen-utilizing organisms.

. . . .

Protein molecules that serve as signaling molecules, transporting information between cells, as well as regulators of nearby cells, via cell-surface receptors.

CYTOKINESIS

. . . .

CYTOKININ

. . . .

CYTOPLASM

C

The division of the cytoplasm into two separate daughter cells from a single eukaryotic cell; this serves to preserve chromosome counts across generations. This process begins following anaphase, when the sister chromatids are separated. Cytokinesis varies in animal cells, plant cells, and bacteria cells.

. . . .

A type of phytohormone (plant hormone) found in plants that facilitates the process of cytokinesis in conjunction with auxin, thereby regulating cell growth, shoot and root morphogenesis, axillary bud growth, apical dominance, and the aging process.

. . . .

In eukaryote cells, the complete inner contents of the cell, not including the nucleus (the contents of the nucleus make up the nucleoplasm), within the cell membrane; in prokaryote cells, which lack a cell nucleus, the cytoplasm refers to the cell's entire contents. The majority of a cell's activities occur within the cytoplasm, including cell division and glycolysis. The cytoplasm is mostly water, and appears transparent and gel-like; it consists of three main parts: the cytosol, the organelles, and the cytoplasmic inclusions.

CYTOSINE

· ·

CYTOSKELETON

· ·

CYTOSOL

C

A compound with the molecular formula $C_4H_5N_3O$ that is a primary base in DNA and RNA (the others are adenine, guanine, and thyamine). It is derived from the heterocyclic organic compound pyrimidine.

· ·

The structural cellular framework found in the cytoplasm of both eukaryotic and prokaryotic cells. Cytoskeletons play key roles in a cell's makeup, shape, and function, and are involved in cellular division and intracellular transport. Cytoskeletons in eukaryotic cells are protein-based and are made up of intermediate filaments, microfilaments, and microtubules.

· ·

The chambered, intracellular fluid within a cell's cytoplasm; it is made up of a variety of substances in multiple organizational levels, including water, small molecules, enzymes, protein complexes, and ions in varying concentrations. The cytosol is involved in multiple cellular processes, including cytokinesis, cellular transport, cell signaling, osmoregulation, and metabolism.

D

DAUGHTER CELL

. . . .

DECIDUOUS

. . . .

DELETION

D

One of the resulting cells following the division of a parent cell (into two or more daughter cells) during mitosis or meiosis in eukaryotic cells and binary fission in prokaryotic cells. A healthy daughter cell is genetically identical to the parent cell and is capable of cell division itself.

. . . .

Term often used to describe various plants (including trees, flowers, and shrubs) that shed certain body structures, often on a seasonal or reproductive cycle, once the part no longer serves a direct functional purpose. This typically includes the shedding of leaves, flower petals, and ripe fruit. The term *deciduous* can also be used in reference to animals, including deer antlers and baby teeth.

. . . .

A genetic abnormality (mutation) in which a portion of a DNA sequence or chromosome is missing, resulting in the loss of key genetic material. Terminal deletion occurs toward the end of a chromosome, and interstitial (intercalary) deletion occurs in the interior of a chromosome. Deletion can be the result of chromosomal translocation, chromosomal inversion, unequal crossing over during mitosis or meiosis, and breakage. Deletions can lead to serious genetic disorders and can even be fatal.

DENITRIFICATION

. . . .

DENSITY-DEPENDENT FACTOR

. . . .

DEOXYRIBONUCLEOTIDE

D

A form of respiration that entails the breaking down and reduction of nitrates, primarily in heterotrophic bacteria, and the release of oxygen (for respiration) and nitrogen. As the nitrogen returns to the atmosphere, the nitrogen cycle is completed.

· ·

A variable or factor that exerts an influence on a population (positive or negative) as the population's density increases or decreases; this can serve to regulate the size of a population to its benefit.

· ·

The molecular unit of DNA, the primary holder of an organism's genetic information; it is comprised of three parts:
  1. a phosphate group
  2. a deoxyribose sugar
  3. a nitrogenous base

DEPENDENT VARIABLE

. . . .

DERMIS

. . . .

DESMOSOME

D

A term used in experimental design and statistical analysis, referring to the condition, event, or factor being observed or measured for change when the independent variable is manipulated.

· ·

A skin layer consisting of connective tissue, mechanoreceptors, sweat glands, hair follicles, sebaceous glands, apocrine glands, lymphatic vessels, and blood vessels. The dermis is located between the epidermis and subcutaneous tissues, and is composed of collagen, elastin, and glycosaminoglycans. It is divided into two layers (the reticular dermis and the papillary region), and serves to provide protection and cushioning to the organism, as well as to facilitate nourishment and waste processing.

· ·

A protein-based junction-like cell structure that is located on the lateral sides of plasma membranes; designed to anchor cells and facilitate cell binding.

DETERMINATE GROWTH

. . . .

DETRITUS

. . . .

DETRIVORE

D

A predetermined biological growth level; once an organism reaches this genetically determined size, it will stop growing.

· ·

Dead organic matter; typically, dead animal or plant tissue or material derived and discarded by living tissue. Detritus is often decomposed by microorganisms and reused within the ecosystem.

· ·

An organism whose primary food and nutrient source is dead organic matter; detrivores serve to make beneficial use of discarded detritus within an ecosystem. Some common detrivores include dung flies, millipedes, slugs, wood lice, sea cucumbers, and various types of worms.

DEUTEROSTOME

. . . .

DIATOM

. . . .

DICOT

D

A classification of enterocoelomate animals that are characterized by coeloms that undergo the process of enterocoely, indeterminate cell cleavage, and blastopores that develop into functioning anuses.

. . . .

One of a group of algae, typically characterized as unicellular with frustules (silica cell walls). Diatoms are commonly found in damp environments (freshwater or oceans) and soil. They are often used to observe environmental conditions and monitor water quality.

. . . .

A type of flowering plant that produces two embryonic cotyledons (seed leaves).

DIFFERENTIAL MIGRATION

. . . .

DIFFUSION

. . . .

DIHYBRID CROSS

D

A specific type of animal migration in which there are some members of a population or species that do not migrate equally, often due to gender or age.

. . . .

A type of naturally occurring biological transport in which substances travel according to the concentration gradient—from areas of greater concentration to areas of lesser concentration.

. . . .

A type of cross-breeding, often used in experiments to determine dominant and recessive genes in two characteristics, wherein two organisms that differ in two distinct, observable character traits are mated.

DIOECIOUS

. . . .

DIPLOID CELL

. . . .

DIRECTIONAL SELECTION

D

A species that possesses separate members with distinct male and female reproductive organs (as opposed to a hermaphroditic species, whose organisms possess both male and female organs); this term is typically used to refer to plant species that produce separate male and female plants.

· ·

A type of cell that contains two chromosome sets, one inherited from each parent (human cells are diploid), as opposed to polyploid cells, which contain more than two sets.

· ·

A type of natural selection wherein the allele frequency for a specific favored phenotype shifts in a favored direction and increases, regardless of relative dominance, so that favored traits become fixed. Directional selection often occurs when a migrating population must adapt to new environmental pressures.

DIURNAL BEHAVIOR

. . . .

DIVERSIFYING SELECTION

. . . .

DIZYGOTIC TWINS

D

Refers to a type of plant or animal whose active behavior primarily occurs during daylight hours, with resting hours occurring at night (as opposed to nocturnal). Humans are diurnal, as are a wide variety of birds, insects, mammals, and reptiles, as well as many plant varieties.

. . . .

Change in the allele frequency distribution, wherein the more extreme values of a trait are favored over intermediate values, potentially leading to trait variance and speciation.

. . . .

Twin organisms (fraternal or non-identical twins) that result from the development of two separate eggs implanted simultaneously in the uterus, which are fertilized separately and form two zygotes.

DNA (DEOXYRIBONUCLEIC ACID)

. . . .

DOMINANT ALLELE

. . . .

DUODENUM

D

An essential macromolecule that contains an organism's genetic information (known as genes) and instructions, which regulate its developmental path. DNA consists of repeating nucleotide units in long polymers, along with phosphate groups and sugars (genetic information is coded as a result of the sequencing of molecular nucleobases on each sugar, and the information is processed via transcription). DNA exists in a double helix structure (structured through hydrogen bonding and base stacking), is housed within the cell nuclei of plants and animals, and is copied and spread through DNA replication during cell division.

. . . .

The allele that is fully recognized during pairing, masking the genetic expression of another allele during pairing. For example, if two different alleles of a gene (Y and Z) are paired, the genetic combinations YY, ZZ, and YZ can result.

- If YZ heterozygotes exhibit the same phenotype as YY, then Y is the dominant allele and Z is recessive.
- If YZ heterozygotes exhibit the same phenotype as ZZ, then Z is the dominant allele and Y is recessive.

. . . .

A structure found in most complex vertebrates; the first (and shortest) section of the small intestine, where the majority of food breakdown (by enzymes) during digestion occurs.

DUPLICATION

. . . .

The copying of an area of DNA containing a gene within a part of a chromosome (or the copying of an entire chromosome). This aberration typically occurs as the result of an error during meiosis between homologous chromosomes that are not properly aligned.

. . . .

ECHINODERM

· ·

ECOLOGICAL EFFICIENCY

· ·

ECTODERM

E

A populous phylum of marine animals that are found within every ocean and are abundant in deep seas. They are typically characterized by their five-point symmetrical shape and their ability to regenerate lost limbs, organs, and tissues. They include sea stars, sea urchins, and sand dollars.

. . . .

The efficiency of energy transference across a level of the food chain (trophic level) that contains all living organisms and highlights their function within the biosphere.

. . . .

Within developing animal embryos, the ectoderm is the outside germ layer (along with the middle [mesoderm] layer and proximal [endoderm] layer). It consists of three main sections—the external ectoderm, the neural crest, and the neural tube. During embryonic development, the ectoderm gives rise to an organism's anus, epidermis, tooth enamel, nervous system, hair, nails, nostrils, sweat glands, and lining of the mouth.

ECTOTHERM

· ·

ELECTROCHEMICAL GRADIENT

· ·

ELECTRONEGATIVITY

E

An animal that does not possess a suitable internal heat source and must utilize external, environmental sources of heat (i.e., sunlight, etc.) to maintain and regulate its body temperature. Such animals include various amphibians, fish, and reptiles. Ectotherms typically live in environments where temperatures are fairly constant, and their behavioral patterns adapt to allow them to maintain a suitable body temperature.

. . . .

The diffusion gradient of an ion, or its gradient of electrochemical potential. This term includes both an electrical potential component (difference in charge across the lipid membrane) and a chemical concentration component (ion concentration difference), and determines the direction of ion movement across a membrane. The term is often used when referring to ATP synthesis.

. . . .

An innate property of an atom or functional group that refers to its tendency to draw electrons to itself; it is largely determined by an atom's nuclear charge and the positioning of electrons on its shell.

EMBRYOGENESIS

· ·

ENDERGONIC REACTION

· ·

ENDOCYTOSIS

E

The development of an embryo into a fetus as a result of the successful fertilization of an ovum (fusion of gametes), creating a zygote.

. . . .

A chemical reaction that involves the absorption of free energy (first-law thermodynamic energy available for systematic work) from its surroundings; this is due to the fact that more energy is needed to initiate the reaction than the amount that results (energy loss).

. . . .

The process of cellular uptake and absorption of important molecules and particulate matter; the result of endocytosis is the creation of an intracellular vesicle through the plasma membrane.

ENDODERM

. . . .

ENDOPLASMIC RETICULUM (ER)

. . . .

ENDORPHINS

The innermost germ cell layer within developing animal embryos (along with the outer ectoderm and the middle mesoderm layers). The endoderm gives rise to the liver, lungs, pancreas, and lining of various organ systems, including the digestive tract.

• •

A network of cells in eukaryotic organisms that form a grouping of cisternae, tubules, and vesicles. Rough endoplasmic reticula are studded with ribosomes that produce proteins; smooth endoplasmic reticula are involved in various metabolic processes and the synthesis of essential lipids; sarcoplasmic reticula store and pump calcium ions.

• •

Hormones that are produced by vertebrates (from the hypothalamus into the spinal cord and brain, and from the pituitary gland into the blood) during intense experiences (exercise, fear, excitement, pain, etc.) that serve to create positive, euphoric feelings and inhibit the perception of pain.

ENDOSKELETON

. . . .

ENDOSYMBIOSIS

. . . .

ENDOTHERM

E

A hard, internal anatomical structure composed of mineralized tissue (bones) within the soft tissue of vertebrates. It serves as a means of structure, support, and protection, and helps facilitate locomotion.

. . . .

A relationship between two living organisms wherein one organism (an endosymbiont) lives inside the body of another organism. This relationship is sometimes necessary for the survival of the host or the endosymbiont. Mitochondria, chloroplasts, and a variety of human parasites are involved in endosymbiosis.

. . . .

Organisms (e.g., mammals) that produce the requisite heat they need to maintain and regulate their body temperatures and survive through internal, homeostatic means including increasing metabolic rates and muscle shivering. Endotherms are much less vulnerable to external, environmental conditions than are ectotherms.

ENZYMES

. . . .

EPINEPHRINE

. . . .

EPISTASIS

E

Protein molecules that act as catalysts in reactions, speeding up the rate of a reaction. Enzymes are substrate-specific (lock-and-key model), so are selective in the types of reactions they are involved with, and serve to lower the activation energy of a reaction, thereby increasing its rate—without being consumed in the process. The utility of an enzyme is affected by inhibitors (which decrease enzyme activity), activators (which increase enzyme activity), environment, temperature, and pressure.

. . . .

Also called adrenaline; a hormone produced in the central nervous system and within the adrenal glands in response to stressors, including exercise, excitement, physical threats, and various unexpected stimuli. Epinephrine binds to various adrenergic receptors in the body and affects almost all body tissue. It serves to regulate blood flow, heart rate, and metabolic activity in an organism; facilitates energy production; and is an essential component of the sympathetic nervous system's fight-or-flight response.

. . . .

The occurrence wherein the expression of a gene is altered by another gene or genes, leading to an epistatic gene whose phenotype is expressed and an altered or suppressed hypostatic gene.

EPITHELIAL TISSUE

....

EPITOPE

....

EQUIVALENCE POINT

AP* BIOLOGY FLASH REVIEW

E

An essential type of animal tissue, made up of sheets of densely packed cells, that serves to line the body's cavities and organs, form various glands, protect underlying body tissue from toxins and trauma, regulate internal chemical exchange and transport, and facilitate glandular hormone secretion.

· ·

The part of an antigen, located on its surface, that is chemically recognized by the paratopes of antibodies in immune systems, thereby eliciting an immune response.

· ·

The point during a chemical reaction (titration) when the amount of a titrant and the substance being titrated are equivalent; also known as the stochiometric point. This determines the minimum amount of a titrant needed to neutralize or react with the substance being titrated. The equivalence point can be determined using a pH indicator, pH meter, potentiometer, or isothermal titration calorimeter; through spectroscopy or amperometry; or by measuring a reaction's conductance, color change, or precipitation.

ETHOLOGY

. . . .

ETHYLENE

. . . .

EUCHROMATIN

E

A subset of zoology; the scientific study of animal behavior patterns and learning (including evolutionary origins and adaptive significance), as well as general behavioral processes.

· ·

A gaseous plant hormone that serves to regulate fruit ripening, flower development and opening, and leaf shedding, and an organic compound that has a wide variety of industrial applications.

· ·

A type of chromatin, which includes the DNA and protein within a cell's nucleus, that is characterized by its loose, unraveled necklace-like structure and heavy concentration of genes. Euchromatin is found in both eukaryotic and prokaryotic cells, and is involved with DNA transcription (to mRNA).

EUKARYOTE

....

EUSOCIAL

....

EUTHERIAN MAMMALS

E

A type of organism with cells that have nuclei containing genetic material (DNA) encased in membranes (as opposed to prokaryotes). Eukaryotes include all animals, plants, and fungi.

. . . .

A specific type of social organization within a population, wherein sterile members of a community perform specialized tasks in support of reproductive members of the community. Various societies of insects, such as bees, ants, and termites, are eusocial.

. . . .

A classification of placental mammals whose offspring develop within a uterus and are connected to their mothers by a placenta. They are typically identifiable by distinct foot, ankle, and jaw features.

EUTROPHICATION

. . . .

EXAPTATION

. . . .

EXERGONIC REACTION

A natural environmental response due to excess pollutants (artificial or natural) being introduced into a body of water, typically from industries and cities. This often leads to a surge in populations of species that feed off the nutrients introduced into the water, including various forms of algae and phytoplankton, as well as a host of negative effects, such as mineral depletion; depletion of essential gases, including oxygen; and decreases in certain other aquatic populations.

. . . .

A structure or trait that has the ability to evolve and shift its function in order to best adapt to changes in environmental conditions.

. . . .

A chemical reaction that results in a decrease in free energy in the final state as compared to the initial state. The ΔG (change in Gibbs free energy) following an exergonic reaction is negative.

EXOCYTOSIS

. . . .

EXON

. . . .

EXTRACELLULAR DIGESTION

E

The process wherein a cell empties the contents of its secretory vesicles (including proteins and lipids) out of the cell membranes for extracellular use. Exocytosis includes vesicle transport, tethering, docking, priming, and fusion.

....

The nucleic acid sequence within a cell that is represented in a mature RNA molecule and codes information for protein synthesis.

....

A type of digestion performed in organisms that digest food externally (saprobiontic organisms), wherein the organisms secrete enzymes onto their food through their cell membranes, which break down the food into digestible molecules prior to absorption.

EXTRACELLULAR MATRIX

. . . .

EXTRAEMBRYONIC MEMBRANES

. . . .

Within animal tissue, the extracellular area, consisting of proteins and polysaccharides, that supports and protects animal cells and helps regulate cell behavior and communication between cells. The extracellular matrix consists of the basement membrane and the interstitial matrix.

. . . .

The membranous layers that are designed to protect and support the embryo, as well as facilitate embryonic growth and development. They include the following:

- The allantois handles embryonic waste and helps facilitate useful gas exchange.
- The amnion protects and cushions the embryo during development.
- The chorion is the layer that covers the other extraembryonic membranes.
- The yolk sac provides nourishment to the developing embryo.

. . . .

F

FATTY ACID

. . . .

FIBRIN

. . . .

FIBROBLAST

F

A carboxylic acid with an even-numbered chain of carbon atoms that vary in length, serve as essential food sources, and yield ATP. Fatty acids can be either saturated (those without double bonds) or unsaturated (those with double bonds).

. . . .

A fibrous blood protein that helps to facilitate the clotting of blood in an organism by providing a meshlike covering over a wound. It is also involved in signal transduction and the activation of blood platelets.

. . . .

Cells that help to facilitate healing in animals as well as provide a structural foundation for connective animal tissue through the release of collagen and adding to the extracellular matrix.

FIXED ACTION PATTERN (FAP)

. . . .

FLAGELLATE

. . . .

FLAGELLUM

F

A hardwired and instinctive behavior pattern in response to an external sensory stimulus. Once a fixed action pattern has begun, it must be fully executed.

. . . .

An organism (plant, animal, or microscopic) or cell (e.g., sperm cell) that is characterized by flagella, which are generally used for propulsion and locomotion.

. . . .

A long, lashlike projection used by some single cells (eukaryotic and prokaryotic) or some single-celled organisms to move with a whiplike motion.

FLUID MOSAIC MODEL

. . . .

FOUNDER EFFECT

. . . .

FRAGMENTATION

F

The model used to describe the structure of cell membranes; in the fluid mosaic model, the membrane of a cell is a liquid layer through which protein molecules and lipids diffuse.

· ·

The process wherein a subset of individuals from a population establishes a new population, which results in genetic drift, the loss of genetic variation, and potential genetic and trait differences from the parent population.

· ·

A process of asexual reproduction in which a portion of an organism can grow into a whole organism; this is used extensively in multicellular invertebrates such as sea stars.

FUNGI

. . . .

F

Organisms organized in a distinct kingdom that have some characteristics of plants and other characteristics that make them more animal-like. They lack chlorophyll and cannot perform photosynthesis, so they don't produce their own food (called heterotrophs). However, they reproduce by spores like plants do.

. . . .

GAMETANGIUM

. . . .

GAMETE

. . . .

GAMETOGENESIS

G

A haploid organ responsible for gamete reproduction in various algae, fungi, protists, and plant gametophytes. Archegonia are female gametangia that produce egg cells; antherdia are male gametangia that produce sperm cells.

· ·

One of the sex cells produced by either male or female organisms that join together during sexual reproduction—sperm in males, egg cells in females.

· ·

The process of cell division (mitotic division or meiotic division) and differentiation within diploid or haploid cells that leads to the formation of gametes.

GANGLION

. . . .

GAP JUNCTION

. . . .

GAS EXCHANGE

A clustered mass of nerve cell bodies (ganglion cells). Ganglia play an important role in the peripheral and central nervous systems of organisms, serving as key relay points and facilitating connections between various neurological structures. Dorsal root (or spinal) ganglia contain nerve cell bodies in afferent spinal nerves. Autonomic ganglia serve as junction points between the autonomic nerves of the central nervous system and target organs.

. . . .

A junction between a pair of animal cells that allows for connection at the cytoplasm, at two hemichannels across the intercellular space, and facilitates the passage of molecules, ions, and currents.

. . . .

In humans, gas exchange between the organism (from the blood) and the external environment (from the air) occurs through the actions of the heart and lungs, through the process of respiration. The alveoli are the sites of oxygen/carbon dioxide diffusion; oxygenated blood is circulated through the body for use.

GASTRIN

. . . .

GASTROPOD

. . . .

GASTRULA

A key digestive hormone (peptide hormone) that is secreted into the bloodstream through the stomach and stimulates gastric juice secretion from the parietal cells, to help facilitate the processing of food. Gastrin release is triggered by stimulus conditions, including stomach distension.

. . . .

A large and diverse class of organisms that are included in the phylum Mollusca, which includes snails and slugs (terrestrial and marine varieties). They are one of the largest taxonomic classes within the biosphere (second only to insects), and there is huge interclass variance in anatomical makeup, behaviors, and environments.

. . . .

In animals, this is a stage of the embryo following cleavage (blastula formation) and consists of a hollow ball of cells. The gastrula contains three layers—the ectoderm, mesoderm, and endoderm; these diverse layers each form different types of tissues and organs within the developing organism.

GENE EXPRESSION

. . . .

GENOME

. . . .

GENOTYPE

G

The process through which a genotype (genetic cellular makeup of a cell or organism) spurs the expression of a phenotype (observable trait/characteristic composite of an organism), through the interpretation of the genetic code within the DNA of an organism. Eukaryotes, prokaryotes, and all known complex biological life forms utilize genetic expression.

· ·

The sum total of an organism's genetic material (genes), encoded into the DNA or RNA, which includes its hereditary information. In humans, this includes the 46 chromosomes within the diploid cells (23 from each of the human haploid gametes, egg and sperm).

· ·

The genetic makeup (allele makeup) of a cell or organism that ultimately determines the physical characteristics (phenotype) of that organism.

GIBBERELLIN

. . . .

GLAND

. . . .

GLIAL CELL

G

A type of plant hormone that serves an essential role in growth and development. Gibberellins trigger stem and leaf growth, and stimulate seed germination and fruit development.

· ·

A structure within organisms, made up of specialized epithelial cells, that creates secretions for release of essential bodily substances, including hormones. Endocrine glands secrete directly into the bloodstream; exocrine glands secrete through ducts into body cavities.

· ·

One of the cells within the nervous system that are nonconducting (do not transmit information via electrical/chemical signaling) and that serve to provide insulation, nourishment (nutrients and oxygen), protection, and support for neurons throughout the brain and nervous system. They also maintain homeostasis and form myelin.

GLUCAGON

. . . .

GLUCOCORTICOID

. . . .

GLUCOSE

G

A key peptide hormone, produced and secreted through the alpha cells of the endocrine portion of the pancreas, that is designed to raise blood glucose levels when blood sugar levels are too low by promoting gluconeogenesis and glycogenolysis. Glucagon acts as a counterbalance to insulin, which serves to lower blood glucose levels, in the maintenance of a healthy and stable blood sugar level.

· ·

A hormone (corticosteroid hormone), produced and secreted through the adrenal cortex, that binds to glucocorticoid receptors and helps regulate glucose metabolism and inhibit overactive immune system function in allergies, asthma, and autoimmune system disorders.

· ·

A simple sugar with the chemical formula $C_6H_{12}O_6$; it is the initial product of photosynthesis and essential in cellular respiration. Glucose is a very common monosaccharide in animal and plant bodies, and is a major source of energy for living organisms.

GLYCOLIPID

. . . .

GLYCOLYSIS

. . . .

GLYCOPROTEIN

A naturally occurring lipid molecule with an organic carbohydrate compound attached. Glycolipids are an important source of energy and facilitate chemical recognition and processing.

· ·

A metabolic pathway in the cells of almost all anaerobic and aerobic organisms, and one of the key energy-producing reactions in organisms. This chemical process entails the splitting of glucose into pyruvate, alongside the release of energy that leads to the development of essential ATP, $FADH_2$, and NADH.

· ·

Integral membrane proteins (IMPs) with covalently attached carbohydrates that play a role in cellular communication.

GOBLET CELLS

. . . .

GOLGI APPARATUS

. . . .

GRADUALISM

Epithelial cells found in the linings of various organs in the intestinal and respiratory tracts that secrete mucin, a protein that forms the basis of mucus, used for lubrication and protection.

· ·

An organelle within eukaryotes that exists near the cell nucleus, whose role is to modify, sort, package, and secrete key macromolecules and proteins.

· ·

The evolutionary theory that states that changes in organisms are the cumulative result of gradual and slow processes.

GRAM STAIN

. . . .

GRANA

. . . .

GRAVITROPISM

G

A method for distinguishing between types of bacteria by the structure of their cell walls and presence of peptidoglycan. Gram-positive bacteria are colored dark blue or violet during Gram staining because of the presence of peptidoglycan in their cell walls, which serves to retain the violet staining material. Gram-negative bacteria do not retain the violet staining dye.

. . . .

Within the chloroplasts of organisms, including plant cells and eukaryotic organisms that conduct photosynthesis, grana are thylakoid stacks within the stroma that are the sites of essential light reactions; photosynthesis occurs on the membranes of thylakoids.

. . . .

The turning direction or direction of growth of an organism (primarily refers to plants and fungi but can also include animals) in response to gravity. Turning or growing toward the direction of gravity is positive gravitropism; turning or growing against the direction of gravity is negative gravitropism.

GROSS PRIMARY PRODUCTIVITY

. . . .

GROUND TISSUE

. . . .

GROWTH FACTOR

G

Within an ecosystem, the total primary production of captured and stored energy across a specified amount of time.

. . . .

A common plant tissue type that includes parenchyma, collenchyma, and sclerenchyma cells, which serve to provide support and fill spaces between the dermal and vascular tissue systems.

. . . .

A specific protein or steroid hormone that is designed to promote the normal growth, development, proliferation, differentiation, and maturation of cells through natural signaling and triggering processes.

GTP (GUANOSINE TRIPHOSPHATE)

. . . .

GUARD CELL

. . . .

GYMNOSPERM

G

A nucloside triphospahte that is a key energy source for substrates in a variety of metabolic reactions, including protein synthesis and glucose production (gluconeogenesis). It can also serve as a substrate for RNA synthesis during transcription, and is also a key component of signal transduction, energy transfer, and genetic translation.

. . . .

One of the specialized plant cells within the leaf epidermis that form the edges of stomatal pathways; guard cells regulate the sizes of the pathways, where water evaporation and gas exchange occurs.

. . . .

Plants that produce unenclosed seeds, on the surface of scales or leaves, that do not form flowers. There are hundreds of existing gymnosperm species, including cypresses, pines, and spruces.

HABITUATION

. . . .

HALF-LIFE

. . . .

HAPLOID CELL

H

A learning process where an animal becomes accustomed to a specific, continuous stimulus, eventually leading to a decrease in resulting behavior elicited. This is a nonassociative learning process in which no reward, punishment, or other stimulus is required.

· ·

The amount of time ($t_{1/2}$) needed for a substance that is decaying at a set rate to lessen by half.

· ·

A cell that contains a single set of chromosomes; humans are diploid organisms, meaning that their cells contain two homologous chromosome sets, one from the mother and one from the father.

HARDY-WEINBERG THEOREM

. . . .

HELPER T CELL

. . . .

HEMOCYANIN

H

The theory that within a specific population, both allele and genotype frequencies are constant and in equilibrium (Hardy-Weinberg equilibrium) across generations, until altered by a significant disturbing influence, including mutations, non-random mating, random genetic drift, gene flow, and so on.

· ·

A type of white blood cell (T_h) that aids in the functioning of an organism's immune system by supporting immune cells—helping B cells produce antibodies and other T cells respond to potentially harmful antigens.

· ·

A respiratory metalloprotein that serves to carry oxygen in the fluid of the circulatory systems (hemolymphs) of various mollusks and arthropods. Their function as oxygen-transporting molecules is similar to that of hemoglobin.

HEMOGLOBIN

. . . .

HEMOLYMPH

. . . .

HETEROCHROMATIN

H

A metalloprotein that consists of iron and serves as the primary molecular oxygen transportation system within the red blood cells of vertebrates, transporting oxygen from the lungs to tissue throughout the body. Hemoglobin also carries carbon dioxide back through the organism to the respiratory system for expulsion.

. . . .

The fluid found in the circulatory systems of mollusks, arthropods, and various other invertebrates, which serves a variety of essential roles, including bathing cells and body tissue.

. . . .

The more tightly packed chromatin type (as opposed to euchromatin) DNA form that helps to facilitate genetic expression in organisms and gene regulation, and to protect the integrity of chromosomes.

HETEROSPOROUS

. . . .

HETEROTROPH

. . . .

HETEROZYGOUS

Among spore-producing plants, a type of vascular plant that produces spores of two different sizes—small microspores that serve the role of male spore and large megaspores that serve the role of female spore (as opposed to homospores, which produce one size and type of spore).

. . . .

An organism (such as an animal) that cannot synthesize its own food and is dependent on consuming organic matter by eating other organisms (either plants or other animals). There are two main types of heterotrophs:
  1. Photoheterotrophs use light for energy and organic compounds such as alcohols, carbohydrates, and fatty acids for nourishment.
  2. Chemoheterotrophs get energy through the oxidation of organic and inorganic molecules in the environment.

. . . .

A condition in diploid organisms where one gene (allele) of a pair is different from the other. Labeling of heterozygous genotypes that do not involve complete single-trait dominance appears in the "Aa" format, where the letter that is capitalized is the dominant allele and the lowercase letter is the recessive allele.

HISTAMINE

. . . .

HISTONE

. . . .

HOMEOSTASIS

H

A substance (organic nitrogen compound) that is released by cells during a body's protective response following an injury, and that serves to trigger a natural inflammatory response, dilate blood vessels, and facilitate greater permeability of capillaries so white blood cells can reach pathogens in infected tissue.

· ·

Any of the small proteins with high levels of alkalinity in eukaryotic cells that serve to package DNA into wound and wrapped nucleosomes, which form the chief repeating units of chromatin. Histones also play a key role in gene regulation.

· ·

The ability of a living organism to maintain a stable internal equilibrium by adjusting its metabolic reactions, typically through the release of hormones in the bloodstream, thereby allowing it to properly function despite fluctuations in its external environment.

HOMOLOGOUS STRUCTURES

· ·

HOMOSPOROUS

· ·

HOMOZYGOUS

H

Structures that exist within two different species due to the fact that they share a common ancestry. The homologous structures need not perform the same function in the two different species.

. . . .

Plants that produce a single type of spore, of an approximately uniform size and shape, that develop both female and male reproductive organs.

. . . .

A condition wherein both genes (alleles) of a pair are the same for a specific trait, such as XX or xx; this also means that the organism's genotype is homozygous.

HORMONE RECEPTOR

. . . .

HUMORAL IMMUNE RESPONSE

. . . .

HYBRIDIZATION

H

Molecules that possess the ability to bind to specific hormones, after which a wave of signaling pathways and targeted chemical reactions is triggered in the body. Hormone receptors are found either on cell surfaces or inside cells, depending on the type of hormone.

. . . .

An immunity response from an organism against bacteria and viruses; it includes the release of antibodies from blood plasma and other body fluids, which bind to antigens of microbes, thus targeting them for attack.

. . . .

The process of breeding among plants or animals that entails crossbreeding two varieties of the same species in an effort to produce offspring that possess certain favorable traits from both varieties.

HYDROGEN BOND

. . . .

HYDROLYSIS

. . . .

HYDROPHILIC

H

A type of electrostatic bond that forms between a hydrogen atom (hydrogen bond donor) and an electronegative atom (hydrogen acceptor). Hydrogen bonds vary greatly in strength, and are responsible for holding together DNA and RNA strands.

· ·

A type of chemical reaction that occurs when a complex molecule is broken down into component parts, following the addition of water, which serves to weaken its bonds. Hydrolysis is a key process in food digestion.

· ·

Having an attraction to or affinity for water; often used in reference to molecules or compounds that typically interact with water and dissolve in water during exposure (salt and sugar are hydrophilic).

HYDROPHOBIC

. . . .

HYPERTONIC

. . . .

HYPHAE

H

Having an aversion to or being repelled from water (fats and oils are examples of hydrophobic molecules).

· ·

Refers to a solution that has a higher concentration of a particular substance on the outside of the cells than within the cells, or a higher solute concentration compared to another substance.

· ·

Threadlike filaments, made of cells encased in tubular cell walls, that branch out and form the mycelium of fungi.

HYPOTHALAMUS

. . . .

HYPOTONIC

. . . .

H

A component of the forebrain in vertebrates, below the thalamus and above the brain stem, that serves to connect and coordinate the endocrine system with the nervous system, manage homeostasis, regulate metabolic activity, and create and secrete neurohormones; it controls key body processes, including temperature, hunger, and sleep.

• •

Refers to a solution that has a lower concentration of a particular substance on the outside of the cells than within the cells, or a lower solute concentration compared to another substance.

• •

IMMUNOGLOBULIN

. . . .

INBREEDING

. . . .

INCLUSIVE FITNESS

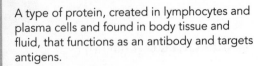

A type of protein, created in lymphocytes and plasma cells and found in body tissue and fluid, that functions as an antibody and targets antigens.

. . . .

The process of plant or animal mating between genetically related organisms. Although this is often used in a controlled environment to enhance the prevalence of desired characteristics, inbred organisms are typically more likely to exhibit defects.

. . . .

The number of offspring (and alleles) that an individual is successfully able to create and pass along through subsequent generations through the reproductive efforts of that individual or of its offspring. Both biological and environmental factors affect an organism's inclusive fitness.

INDEPENDENT VARIABLE

. . . .

INDETERMINATE GROWTH

. . . .

INDUCED FIT

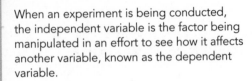

When an experiment is being conducted, the independent variable is the factor being manipulated in an effort to see how it affects another variable, known as the dependent variable.

. . . .

Growth within plants that does not stop once a level of predetermined genetic structural development is reached; in animals, indeterminate growth typically refers to quick growth during the initial stages of life of the organism, with continued, albeit slower, growth throughout adulthood.

. . . .

Changes that occur within an enzyme's active site in an effort to improve the binding strength between the enzyme and a substrate (differential binding or uniform binding). These conformational changes are affected by both internal and external conditions.

INFLAMMATORY RESPONSE

. . . .

INHIBITOR

. . . .

INNATE RELEASING MECHANISM

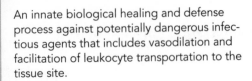

An innate biological healing and defense process against potentially dangerous infectious agents that includes vasodilation and facilitation of leukocyte transportation to the tissue site.

. . . .

An enzyme-inhibiting molecule that binds to target enzymes and functions to decrease or block their activity. This process may be temporary or permanent, can occur through a variety of biological reaction mechanisms, and can serve a variety of functions.

. . . .

A circuitous neurological pathway within the brains of animals that is designed to respond to an external sensory stimulus through an instinctive, hardwired fixed action pattern (FAP).

INNER CELL MASS

. . . .

INSULIN

. . . .

INTERFERON

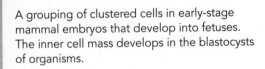

A grouping of clustered cells in early-stage mammal embryos that develop into fetuses. The inner cell mass develops in the blastocysts of organisms.

· ·

A hormone in vertebrate organisms, produced in the pancreas, that serves to lower glucose levels in the blood by promoting the uptake of glucose from the blood to cells throughout the body, including fat and muscle tissues, and into the liver, where it is stored as glycogen.

· ·

A type of protein, produced and released by cells that are hosting tumors or have been infected by pathogens (e.g., bacteria, viruses, or parasites), that triggers defense responses from the immune system to fight the pathogen; this includes activating immune cells, increasing host cell recognition, and helping uninfected cells resist infection.

INTERKINESIS

....

INTERNODE

....

INTERPHASE

I

During cell division (meiosis), interkinesis is a period of intermediate cell rest between meiosis I and II.

· ·

The area on the stems of plants that separates the nodes, which contain the buds that develop into leaves, flowers, cones, roots, or additional stems.

· ·

The period of the cell cycle in which the cell is actively preparing for division (mitosis or meiosis). During interphase, cell size increases, DNA and chromosomes are copied, and essential nutrients are obtained in preparation for division, but no actual division occurs during this phase. Cells spend the majority of their time in this phase.

INTERSPECIFIC COMPETITION

. . . .

INTERSTITIAL FLUID

. . . .

INTERTIDAL ZONE

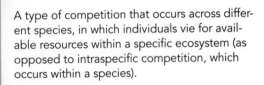

A type of competition that occurs across different species, in which individuals vie for available resources within a specific ecosystem (as opposed to intraspecific competition, which occurs within a species).

. . . .

The primary component of extracellular fluid, found in the interstitial spaces and made up of water, amino acids, fatty acids, sugars, hormones, salts, neurotransmitters, and co-enzymes. Interstitial fluid protects, surrounds, bathes, and transports nutrients and waste in the cells of animals.

. . . .

The area at an ocean's edge between set tide markers so it is visible from land during low tide and underwater during high tide; it is often home to many water-dwelling species despite being an extremely volatile habitat, with constantly changing environmental conditions.

INTRACELLULAR DIGESTION

. . . .

INTRASPECIFIC COMPETITION

. . . .

INTROGRESSION

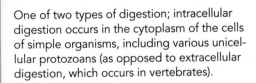

One of two types of digestion; intracellular digestion occurs in the cytoplasm of the cells of simple organisms, including various unicellular protozoans (as opposed to extracellular digestion, which occurs in vertebrates).

· ·

A type of competition that occurs within a species, in which individuals vie for available resources within a specific ecosystem (as opposed to interspecific competition, which occurs across different species).

· ·

The transportation of genes from one species to another as the result of repeated backcrossing of a fertile interspecific hybrid with one of its parent's species.

INVERSION

. . . .

INVERTEBRATE

. . . .

IN VITRO

A chromosomal aberration that occurs when a fragment of a chromosome is reversed (often following breakage and rearrangement), leading to a rearranging of the linear gene sequence. Typically, inversions that are balanced and are not missing any genetic information do not lead to abnormalities.

. . . .

An organism that lacks a backbone or spinal column. The vast majority of animal species (over 95%) are invertebrates; examples include insects, crayfish, lobsters, clams, sponges, jellyfish, and sea stars.

. . . .

Refers to experiments that entail examining a portion of an organism that has been separated from the whole, intact organism, in an effort to get a targeted and focused analysis (often called test tube experiments).

IN VIVO

. . . .

ION

. . . .

IONIC BOND

Refers to experiments that include an entire living organism, as opposed to a dead organism or an isolated potion of an organism (*in vitro*), in an effort to understand how the entirety of a living organism is affected by certain variables; examples include animal testing and human clinical trials.

. . . .

An atom that has acquired an electrical charge by gaining one or more electrons (resulting in a negative charge) or losing one or more electrons (resulting in a positive charge). Monatomic ions are made up of a single atom; polyatomic ions are made up of multiple atoms.

. . . .

A chemical bond between two ions with opposite charges—a cation (an ion with a positive charge) and an anion (an ion with a negative charge). The ionic strength of a bond is determined by the difference in electronegativity between the atoms.

ISOMERS

. . . .

ISOTONIC SOLUTIONS

. . . .

ISOTOPES

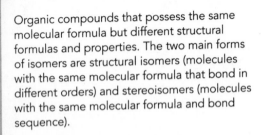

Organic compounds that possess the same molecular formula but different structural formulas and properties. The two main forms of isomers are structural isomers (molecules with the same molecular formula that bond in different orders) and stereoisomers (molecules with the same molecular formula and bond sequence).

. . . .

Solutions that, when compared, have equal solute concentrations.

. . . .

Variations of an element that have different numbers of neutrons and different atomic masses.

KARYOGAMY

· ·

KARYOTYPE

· ·

KERATIN

During bacterial conjugation and fungi reproduction, the fusion of the pronuclei of two cells
(the nuclei of an egg and a sperm cell during
fertilization, prior to fusion).

. . . .

An organizational classification system for
chromosomes within eukaryotic cells, which
takes into account the number, type, and size
of chromosomes in a cell's nucleus.

. . . .

A type of strong, fibrous protein that serves as
a foundational component in various organic
structures, including human skin, hair, and
nails, as well as in animal claws and feathers.

KEYSTONE SPECIES

. . . .

KINESIS

. . . .

KINETIC ENERGY

A term for a species of great influence on and importance in its environment, and the maintenance of species diversity within it. The balance of an ecological community depends on the health and proliferation of its keystone species.

. . . .

A fluctuation in an organism's nondirectional activity in response to a stimulus. Orthokinesis occurs when the intensity of a stimulus affects its activity speed. Klinokinesis occurs when the frequency of the resultant activity is affected by the intensity of the stimulus.

. . . .

The energy of motion of an object; the amount of work that is required to accelerate an object with a defined mass from rest to a specific velocity. Kinetic energy is constant until the object's speed decreases, and an equal amount of energy is required to bring the object from its initial speed to a rest state.

KINETOCHORE

. . . .

K

An area on the centromeres of eukaryote cells, made of protein, where the spindles of sister chromatids attach during cell division (mitosis and meiosis).

. . . .

LAGGING STRAND

· ·

LAW OF INDEPENDENT ASSORTMENT

· ·

LAW OF SEGREGATION

L

A strand of template DNA that stretches away from the replication fork and is discontinuously synthesized.

· ·

A law that states that during gamete formation, gene pairs for separate traits pass independently of each other from parents to their offspring; a singular biological selection does not occur across all traits and gene pairs.

· ·

Law stating that during gamete formation, each parent supplies its offspring with a randomly selected allele copy that forms an allele pair for any given trait; the dominant allele is the key factor in how that trait is expressed.

LEUKOCYTE

. . . .

LIGAMENT

. . . .

LIGAND

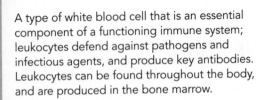

A type of white blood cell that is an essential component of a functioning immune system; leukocytes defend against pathogens and infectious agents, and produce key antibodies. Leukocytes can be found throughout the body, and are produced in the bone marrow.

. . . .

Tough, fibrous connective tissue made of collagenous fibers that connects bones together at a joint.

. . . .

A molecule that forms a coordination complex through binding with a central atom (usually metal) at its receptor site, typically involving donating electrons to the metal.

LIGHT REACTIONS

. . . .

LIGNIN

. . . .

LINKAGE

L

The initial, critical stages of photosynthesis, wherein solar energy is captured by plants and converted to usable forms of chemical energy (ATP and NADPH).

. . . .

An organic polymer typically found in plant cell walls and wood, where it provides structure and support and facilitates internal water transport. It also plays a key role in the ecosystem's carbon cycle.

. . . .

During cell division (meiosis), the tendency for closely located genes (on the same chromosome) to get inherited together. Genetic linkage during crossover is stronger for genes that are located in close proximity.

LINKED GENES

. . . .

LIPID

. . . .

LOOP OF HENLE

L

Genes that are located on the same chromosome and that have a propensity for getting inherited together during cell division.

· ·

A water-insoluble molecular compound that serves to store energy, provide structure to cell membranes, and perform biological signaling during cellular communication. The lipid family includes diglycerides, fats, monoglycerides, phospholipids, triglycerides, and waxes.

· ·

A looping anatomical structure in the kidney, named after the physician Friedrich Gustav Jakob Henle, that regulates an optimal water/salt concentration gradient through reabsorption.

LUMEN

. . . .

LUTEINIZING HORMONE

. . . .

LYMPH

Within any tubular structure in the body (e.g., arteries or blood vessels), the lumen is the inside central cavity.

. . . .

A key protein hormone that plays a role in facilitating reproductive functioning—stimulating ovulation and hormone production processes in females and testosterone production in males. The luteinizing hormone is created in the anterior pituitary gland.

. . . .

The colorless fluid within the lymphatic system, made up of white blood cells, that facilitates transport and waste removal through circulation. It is derived from interstitial fluid.

LYMPHATIC SYSTEM

. . . .

LYMPHNODE

. . . .

LYMPHOCYTE

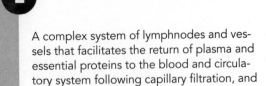

A complex system of lymphnodes and vessels that facilitates the return of plasma and essential proteins to the blood and circulatory system following capillary filtration, and supports normal functioning of an organism's immune system.

· ·

Oval structures made up of compartmentalized spongy tissue that serve a variety of key immune system functions, including removing waste and dead cells through circulation, filtering foreign particles, and facilitating the targeting of potentially dangerous antigens.

· ·

A type of white blood cell within the immune system; lymphocytes include B cells (developed in the bone marrow and facilitate humoral immunity), T cells (developed in the thymus and facilitate cell-mediated immunity), and natural killer cells (defend against tumors and viral infection).

LYMPHOKINE

. . . .

LYSIS

. . . .

LYSOSOME

A type of signaling molecule produced by T cells that facilitates immune system response and the development of antibodies.

. . . .

The process of cell breakdown, typically resulting from enzyme or virus activity or from osmosis.

. . . .

One of the cellular structures within the cytoplasm that house hydrolytic enzymes within a membrane-enclosed area; lysosomes are responsible for cellular digestion and waste processing.

LYSOZYME

. . . .

LYTIC CYCLE

. . . .

A type of enzyme (glycoside hydrolase) that is present in a variety of secretions, including human milk, mucus, saliva, sweat, and tears, and that serves to break down the cellular walls of bacteria.

· ·

A multi-stage cycle of viral reproduction (and the main mechanism for viral replication in organisms), in which the viral DNA of a new phage is a separate molecule within the bacterial cell.

· ·

MACROMOLECULE

. . . .

MACRONUCLEUS

. . . .

MACROPHAGE

One of the large molecules, some of which are the key structural components of living things, including carbohydrates, lipids, nucleic acids, and proteins. Macromolecules are formed through the joining of smaller molecules, often through polymerization.

. . . .

Polyploid (possessing more than two paired sets of chromosomes) structures in ciliates that regulate and control all of the nonreproductive cell processes; reproduction is controlled by a diploid micronucleus.

. . . .

A white blood cell (phagocyte) involved in both adaptive and innate immune system function; macrophages serve to digest and remove cell waste, and trigger an immune response to destroy pathogens and potentially dangerous foreign substances.

MARSUPIAL

. . . .

MAST CELL

. . . .

MECHANORECEPTOR

A classification of mammal whose offspring are born following short gestation periods and finish their development inside a maternal pouch (marsupium). Marsupials include kangaroos, koalas, opossums, and wombats.

. . . .

A type of cell containing histamine and heparin, which helps prepare immune system reactions to allergens and parasites, and facilitates innate immune system function, the inflammatory process, and tissue repair.

. . . .

A specialized type of sensory receptor in organisms that is designed to respond to mechanical pressure, displacement, and distortion.

MEIOSIS

. . . .

MEMBRANE POTENTIAL

. . . .

MERISTEM

M

A type of cell division that occurs within eukaryotic cells that reproduce sexually, leading to the creation of gametes (sperm or egg cells) or spores with a combination of maternal and paternal DNA.

. . . .

The difference in charge (electrical potential) between a cell's interior (cytoplasm) and exterior (extracellular fluid) due to differing ion concentration gradients, stimulating transmembrane moment and activity.

. . . .

A type of plant tissue made up of tightly packed, undifferentiated cells that remain active throughout a plant's life; it is where development and growth of the organs (e.g., leaves and flowers) take place.

MESODERM

. . . .

MESOPHYLL

. . . .

MESSENGER RNA (mRNA)

An embryonic layer of germ cells that develops into the muscle, bone, connective tissue, coelom lining, and part of the reproductive system during growth. The mesoderm is one of three primary germ cell layers, along with the (inner) endoderm and (outer) ectoderm.

. . . .

Within plants, the mesophyll is the middle ground tissue of the leaves, between the upper and lower epidermis layers, where photosynthesis takes place.

. . . .

An RNA molecule that delivers specific protein-making coding instructions from DNA to the ribosomes of the cytoplasm, where protein synthesis occurs.

METABOLIC PATHWAYS

. . . .

METABOLISM

. . . .

METAPOPULATION

M

A series of essential biochemical reactions that take place within cells, in which chemicals are modified and form new products that are either used immediately, stored for later use, or used to initiate a new metabolic pathway. Within a cell, the sum total of metabolic pathways makes up its metabolic network.

. . . .

Within an organism, the complete set of catabolic (metabolic pathways that create and release energy by breaking down molecules) and anabolic (metabolic pathways that build new molecules, which requires energy) chemical processes and reactions, which allow the essential functions of sustaining life to occur.

. . . .

Within a species, a metapopulation is a subdivided, fragmented group of separate and distinct populations in relation to the overall population.

METASTASIS

. . . .

METHYLENE BLUE

. . . .

MICROBE

Within an organism, the spread of a disease or infection from one part of the body (the origin) to another area; this term typically is used to refer to malignant cancer cells and infections.

. . . .

A chemical compound with the molecular formula $C_{16}H_{18}N_3SCl$ that often is used as a staining agent during cell analysis, often to examine DNA or RNA, as well as for water testing and sulfide analysis.

. . . .

A very small organism that cannot be seen with the unaided eye and requires the use of a microscope or at least a magnifying lens. We can also detect microorganisms by chemical tests; these living beings are everywhere, even in extreme environments such as very hot springs, very cold and dry areas, and even deep in the ocean under tremendous pressure. Some of these organisms cause diseases in animals, plants, and humans; however, most are beneficial to us and the earth's ecosystems. In fact, we are utterly dependent upon microbes for our quality of life.

MICROFILAMENT

. . . .

MICRONUCLEUS

. . . .

MICRONUTRIENT

Within eukaryotic cells, the thin yet flexible and strong filaments made of actin (a protein) that make up the cytoskeleton of the cytoplasm. Microfilaments help facilitate cell motility and contraction.

. . . .

The tiny cell nucleus in simple ciliate protozoans that undergo mitosis and conjugation to multiply.

. . . .

Any of the inorganic nutrients required by living organisms in small amounts for sustained health and proper functioning. Micronutrients include a variety of vitamins and key trace minerals such as cobalt, chromium, copper, iodine, iron, and manganese.

MICROSPORES

. . . .

MICROTUBULE

. . . .

MICROVILLI

Plant spores that turn into male gametophytes. In heterosporous plants, microspores combine with megaspores (which turn into female gametophytes); in seed plants, microspores become pollen grains.

. . . .

Hollow cylindrical rods made of a protein called tubulin that are found in cell cytoplasms. Microtubules provide structure and support to eukaryotic cells, and facilitate cellular transport.

. . . .

Tiny projections on the membranes of various cells, which serve to increase cellular surface area and facilitate essential cell processes, including absorption, adhesion, secretion, and transduction.

MIDDLE LAMELLA

. . . .

MIMICRY

. . . .

MINERALOCORTICOID

Within plants, a layer of sticky extracellular material (mostly pectin) between primary walls that serves to cement adjoining cell walls, adding stability.

. . . .

A natural phenomenon in which a species shares a similarity (e.g., appearance, behavior, sound, or scent) to another species, and receives some benefit from that similarity. This often occurs as a result of selective evolution.

. . . .

A type of steroid hormone that helps regulate the levels of water and salt in an organism by facilitating reabsorption of water and sodium and secretion of potassium for homeostasis.

MITOCHONDRION

....

MITOSIS

....

MITOTIC SPINDLE

In most eukaryotic cells, an organelle that serves to generate cellular energy (ATP) and facilitate a variety of key processes, including cell signaling, differentiation, and growth.

. . . .

A division process in eukaryotic cells in which cells divide through cytokinesis to produce two daughter cells—and the chromosomes within a cell's nucleus split into two identical yet separate sets. The stages of mitosis include prophase, prometaphase, metaphase, anaphase, and telophase.

. . . .

A subcellular structure made up of proteins and spindle microtubules that plays a key role during mitosis, separating the chromosomes across daughter cells during division.

MOLARITY

. . . .

MOLE

. . . .

MONOCOT

Also known as molar concentration, molarity is the amount (concentration) of a solute (number of moles) in a solution.

· ·

A chemical unit of measurement; the amount of a chemical substance that equals Avogadro's number of molecules ($6.02214179 \times 10^{23}$).

· ·

A type of flowering plant that produces a single seed leaf (cotyledon), as opposed to dicots, which produce two cotyledons. Monocots include orchids, lilies, daffodils, tulips, rice, wheat, sugar cane, and bamboo.

MONOCYTE

. . . .

MONOECIOUS

. . . .

MONOHYBRID CROSS

A type of white blood cell produced in the immune system of vertebrates. Monocytes serve a variety of key roles within the immune system, including production of macrophages used for innate and adaptive immunity, and production of dendritic cells for antigen synthesis.

. . . .

A species of plant that possesses both male (staminate) and female (carpellate) flowers on a single plant. Examples of monoecious plants include birch trees, pine trees, fig trees, and corn.

. . . .

A type of experimental breeding that utilizes two individuals that possess different genetic traits (alleles) for a specific character at a specific genetic locus, and are homozygous for that trait.

MONOMER

. . . .

MONOSACCHARIDES

. . . .

MONOSOMY

M

A type of molecule that has the ability to bind to other molecules and form polymers; monomers can be organic (e.g., glucose) or synthetic (e.g., vinyl chloride).

. . . .

The simplest unit of carbohydrate, also referred to as simple sugars. Monosaccharides are important building blocks for essential disaccharides (e.g., sucrose) and polysaccharides (e.g., cellulose), and include fructose and glucose.

. . . .

A type of chromosomal aberration in which one chromosome is missing from a normal diploid pair, leading to various adverse clinical syndromes and impacted development and growth.

MONOTREME

· ·

MONOZYGOTIC

· ·

MORPHOGENESIS

A rare type of mammal that possesses the ability to lay eggs, rather than giving birth to developed, live offspring. Monotremes include the platypus and a few types of anteaters.

. . . .

Refers to a type of biological twin who develops when a single zygote is formed following the fertilization of one egg; during development, the zygote splits into two distinct embryos.

. . . .

A broad term used to define the set of biological processes that lead to an organism developing its unique shape, from embryonic development into adulthood. Morphogenetic responses can be the result of internal factors such as hormones or environmental responses such as toxic or radioactive chemical exposure leading to changes.

MORPHOLOGY

. . . .

MOSAIC EVOLUTION

. . . .

MUTAGEN

M

The study of organisms' various structures, forms, and features—including internal structures and functioning (anatomy) and external appearance (eidonomy).

· ·

An evolutionary concept that states that the changes and development of different features and parts of an organism can occur at different times and rates.

· ·

Any chemical or physical agent that has the ability to interact with and alter an organism's genetic material and potentially cause mutations.

MUTATION

. . . .

MUTUALISM

. . . .

MYCELIUM

M

An atypical alteration in the genome sequence of an organism (DNA or RNA) that can lead to large and small genetic changes. Mutations can result from an error during cell division or embryonic development, as well as from chemical exposure, radiation, and viruses.

• •

A relationship between two organisms in which both derive some benefit from the arrangement; mutualism is a type of symbiotic relationship.

• •

A large mass of interconnected, branching hyphae that constitutes the main body of multicellular fungi. However, the mycelium is usually not seen because it is hidden throughout the food source being consumed.

MYCORRHIZA

· ·

MYELIN

· ·

MYOFIBRIL

M

A type of symbiosis (mutualistic relationship) between vascular plant roots and fungi; the relationship provides the fungi with needed carbohydrates, and helps plants access needed phosphate and improve their ability to absorb water and nutrients.

· ·

A type of insulating material, primarily composed of water, lipids, and protein, that makes up the myelin sheath, which insulates the axons of neurons and serves to facilitate the proper transmission and speed of electrical nerve impulses.

· ·

An elongated building block of muscle, composed of various proteins, including thin filaments of actin and thick myosin filaments.

MYOFILAMENT

. . . .

MYOGLOBIN

. . . .

MYOSIN

M

Within the various muscle types, filaments (made up of myofibril) that provide structure and form; myofilaments can be thick, thin, or elastic, and the position and arrangement of myofilaments within muscle determine whether the muscle is obliquely striated, smooth, or striated.

. . . .

A type of protein found in the muscles of vertebrates, which stores and binds to oxygen for strength and support.

. . . .

A type of protein found in muscles that works in conjunction with actin to produce muscle contraction.

MYRISTOYLATION

. . . .

A process of protein modification that occurs in various types of organisms and is involved in apoptosis, cell regulation, signal transduction, and translocation.

• •

NADP

. . . .

NATURAL KILLER CELLS

. . . .

NEGATIVE CONTROL

A type of coenzyme (nicotinamide adenine dinucleotide phosphate) in plants and animals that is instrumental in the synthesis of free radicals, lipids, cholesterol, and nucleic acids, and in various anabolic reactions. In plants, NADP is involved in the process of photosynthesis.

. . . .

A type of quick-acting defensive immune system cell that responds to and attacks cells infected by viruses or tumors, even in the absence of antibody activation. Natural killer cells (NK cells) are involved in both innate and adaptive immune system responses.

. . . .

Within controlled scientific experiments, this refers to a control group in which there is no expectation of observable change or phenomenon; no effect is anticipated as a result of experimental manipulation.

NEMATOCYST

. . . .

NEPHRIDIUM

. . . .

NEPHRON

Within organisms in the Cnidaria phylum, which includes sea anemones, corals, and jellyfish, a stinger-like structure that is used for defense against predatory creatures and for capture of prey.

· ·

In invertebrate organisms, a structure that is involved in the excretory process—removing waste products for healthy functioning. Nephridium functioning is similar to that of kidneys.

· ·

In vertebrates, the foundational structural component of the kidneys. Nephrons are regulated by hormones in the endocrine system and serve to filter blood and regulate blood volume, pH, and pressure; regulate electrolyte, metabolite, sodium, and water levels through reabsorption and excretion; and eliminate waste products.

NERITIC ZONE

....

NET PRIMARY PRODUCTIVITY

....

NEURAL CREST

The shallow part of the ocean, also known as the coastal waters, which is characterized by temperate waters with high oxygen and saline levels and abundant aquatic life.

· ·

The amount of total stored and available chemical energy within a defined ecosystem, minus the energy utilized for cellular respiration by the producers of energy.

· ·

A group of cells in developing vertebrate embryos that migrate and form the structural components of smooth muscle, skin pigment cells, craniofacial cartilage and bone, the teeth, and parts of the peripheral nervous system.

NEURAL GROOVE

. . . .

NEURAL PLATE

. . . .

NEURAL TUBE

In developing vertebrate embryos, a longitudinal groove sandwiched between a pair of neural folds; the neural grove deepens and the neural folds stretch, turning into the neural tube, which forms the structural components of the nervous system.

· ·

In developing vertebrate embryos, a thick strip of dorsal ectoderm that helps give rise to the neural tube (a key structural component of the central nervous system) during primary neurulation.

· ·

In developing vertebrate embryos, the neural tube serves as a key foundational component of an organism's central nervous system. The neural tube develops during the primary and secondary neurulation processes into four subdivisions: the prosencephalon (the forebrain, optic vessels, and hypothalamus); the mesencephalon (midbrain); the rhombencephalon (cerebellum and pons); and the spinal cord.

NEUROHORMONE

· ·

NEUROMODULATOR

· ·

NEUROMUSCULAR JUNCTION

A blanket term referring to any hormone that is synthesized and secreted by any of the neurons in an organism. Neurohormones are critical components of a variety of key processes; examples include thyrotropin-releasing hormone TRH, ADH, and oxytocin.

. . . .

A substance (neurotransmitter) secreted by neurons that affects a variety of essential processes within the central nervous system, including regulating brain activity.

. . . .

The space between a motor neuron's branched axon terminal and a muscle fiber; the area where neurons signal muscles to contract.

NEURON

. . . .

NEUROSECRETORY CELLS

. . . .

NEUROTRANSMITTER

A nervous system cell that is essential to nervous system functioning; neurons conduct and transmit electrical and chemical impulses, thereby communicating key information across neural networks. Neurons are often specialized for various tasks, including sensory neurons (sensory system processing), motor neurons (muscle and gland functioning), and interneurons (connecting neurons across the nervous system).

. . . .

Cells within the hypothalamus that receive signal input from neurotransmitters and respond by releasing appropriate hormones into the bloodstream, thus facilitating neuroendocrine integration.

. . . .

A type of chemical messenger that communicates signals from a neuron, across a synapse (synaptic cleft), and to the receptor of a target cell that it binds to. The action of a neurotransmitter follows an action potential or graded electrical potential. Neurotransmitters include amino acids, monoamines, and peptides.

NEUTRAL VARIATION

. . . .

NITRIFICATION

. . . .

NITROGEN CYCLE

A type of DNA variance (genetic diversity) that confers a complete lack of selective genetic advantage.

. . . .

An important nitrogen cycle process in the biosphere; the process in which ammonia is oxidized and transforms into nitrites and then into nitrates.

. . . .

The biological, chemical, and physical processes (including fixation, mineralization, nitrification, and denitrification) of nitrogen transformation, circulation, and utilization within the biosphere.

NITROGENASE

. . . .

NITROGEN FIXATION

. . . .

NODE

A type of enzyme that reduces nitrogen gas in the atmosphere to ammonia.

· ·

A chemical conversion process in which nitrogen from the atmosphere is transformed into ammonia and various nitrogen compounds for a variety of uses. It is an essential, life-sustaining process for all types of life forms and has a variety of industrial uses as well.

· ·

A structural point on the stem of a plant where cones, leaves, flowers, and roots grow. Stems connect to each other on plants at the nodes.

NODES OF RANVIER

....

NONDISJUNCTION

....

NONPOLAR MOLECULE

N

A structural component of neurons, the nodes of Ranvier are the junction gaps between connected segments of myelin sheath. The axon is uninsulated at the nodes of Ranvier; thus, it is where electrical activity at voltage-gated ion channels occurs.

. . . .

A breakdown during the cell division (mitosis or meiosis), wherein homologous chromosome pairs or sister chromatids fail to properly move apart and separate, thereby creating a cell with a chromosome imbalance. This can lead to a variety of developmental syndromes and potential defects.

. . . .

A type of molecule that consists of neutral bonds with no polarity, due to either an equal sharing between atoms of available electrons or a symmetrical polar bond arrangment.

NONSENSE MUTATION

. . . .

NORADRENALINE

. . . .

NOTOCHORD

A type of DNA sequence point mutation in which an amino acid codon transforms into a premature stop codon and leads to a small, truncated, ineffective protein; a variety of genetic disorders can result.

· ·

An important hormone and neurotransmitter produced by the adrenal medulla that serves a variety of nonadrenergic roles, including regulating the fight-or-flight response, raising blood pressure and heart rate, and increasing blood glucose concentration, muscle strength, and fatigue resistance. Also known as epinephrine, its role as a neurotransmitter is important to the central and sympathetic nervous system.

· ·

A long, flexible rod made up of mesoderm cells that is found in chordate embryos and provides embryonic structure and support.

NUCLEAR ENVELOPE

. . . .

NUCLEIC ACID

. . . .

NUCLEOID

A porous membrane found in eukaryotes, made up of a double lipid bilayer, that encases a cell's genetic material, separates the cell nucleus from the cytoplasm, and manages the exchange of key materials from both sides.

· ·

An organic molecule found in all living cells and viruses; nucleic acids in the form of molecules called deoxyribonucleic acid (DNA) and ribonucleic acid (RNA) control cellular functions and heredity.

· ·

In prokaryotic cells, the dense region containing the majority of the genetic material.

NUCLEOLUS

....

NUCLEOSOME

....

Within the cell nucleus, a structure made up of chromosomes that plays a key role in the transcription and assembly of ribosomal ribonucleic acid (rRNA).

. . . .

The basic, beadlike unit of DNA packaging in eukaryotes, consisting of a segment of DNA wound around a protein core composed of two copies of each of four types of histone.

. . . .

OBLIGATE AEROBE

. . . .

OBLIGATE ANAEROBE

. . . .

OMNIVORE

An organism that needs to breathe oxygen for survival, energy synthesis, and growth and development.

· ·

An organism that does not rely on oxygen respiration for survival and growth; certain obligate anaerobes are even harmed or killed by oxygen.

· ·

An organism that consumes both plants and animals as part of its regular food sources; the systems of most omnivores rely on a suitable mix of both edible plant and animal materials in order to maintain optimal health.

OOCYTE

. . . .

OOGENESIS

. . . .

OPERANT CONDITIONING

A female eukaryotic germ cell (gametocyte) that is grown in the ovary and develops into an ovum during meiosis.

· ·

The creation and development of the mature female ovum; the process includes several stages and is the female equivalent of gametogenesis.

· ·

A form of associative learning in which the formation of an organism's voluntary behaviors is shaped by consequences through trial and error. Behaviors are reinforced or avoided based on resultant rewards or punishments.

OPERON

. . . .

OPSONIN

. . . .

ORGANELLE

A unit of functioning genomic DNA in certain prokaryotes and eukaryotes that consists of a cluster of structural genes regulated together by a nucleotide sequence known as a promoter.

· ·

A molecule that targets antigens during an immune response; opsonins also trigger the complement system and serve as binding enhancers during phagocytosis.

· ·

One of the specialized structural subunits within cells (prokaryotic and eukaryotic cells), each serving a specific key function. Organelles within eukaryotic cells include the cytoskeleton, Golgi apparatus, lysosome, mitochondria, nucleolus, nucleus, ribosome, and vesicle. Organelles within prokaryotic cells include the chlorosome, flagellum, nucleoid, and ribosome.

ORGANIC COMPOUND

....

ORGANOGENESIS

....

ORIGIN OF REPLICATION

Gaseous, liquid, or solid chemical compounds in which one or more atoms of carbon are covalently linked to atoms of other elements, such as hydrogen, oxygen, or nitrogen. Examples of organic compounds are carbohydrates, lipids, proteins, and nucleic acids. A few carbon-containing compounds such as carbides, carbonates, and cyanides are considered inorganic.

. . . .

The formation of internal organs through a series of organized integrated processes that transform an amorphous mass of cells into a complete organ in the developing embryo. Organogenesis occurs within the third week through the eighth week of gestation. Organogenesis is also known as *organogeny*.

. . . .

A particular sequence of DNA at which replication is initiated. DNA can be replicated in living organisms such as prokaryotes and eukaryotes, or in viruses. A single origin is sufficient for small DNAs, whereas larger DNAs have multiple origins, and DNA replication is initiated at all of them. Having multiple origins of replication helps to speed up the duplication of genetic material. Origin of replication is also known as the *replication origin*.

OSMOCONFORMER

. . . .

OSMOREGULATION

. . . .

OSMOREGULATOR

An organism whose body fluid solute concentration changes directly with a change in the solute concentrations of the surrounding medium in which the organism lives. Most marine invertebrates are osmoconformers.

. . . .

The process of regulating the levels of water and mineral salts (electrolytes) in the blood. It is a homeostatic mechanism that keeps an organism's fluids from becoming too diluted or too concentrated.

. . . .

An organism that can tightly regulate its body fluid solute concentration irrespective of changes in the surrounding medium in which the organism lives. Almost all aquatic animals are osmoregulators.

OSMOSIS

. . . .

OSTEOCYTE

. . . .

The net movement of solvent molecules through a semipermeable membrane into a region of higher solute concentration. Osmosis occurs in order to equalize the solute concentrations on either side of the membrane. The membrane is permeable to the solvent, but not the solute.

. . . .

A star-shaped cell that lies within the substance of fully formed bone. Osteocytes are the most numerous cells found in mature bone, and can live as long as the organism itself. Each osteocyte occupies a small, round cavity called a lacuna, and has thin, cytoplasmic branches. Osteocytes extend processes through small channels called canaliculi to connect to neighboring cells by means of gap junctions. Nutrients and waste products are exchanged through the canaliculi. Osteocytes originate from osteoblasts (bone-forming cells that actively produce bone tissue).

. . . .

OXALOACETATE

. . . .

OXIDATION

. . . .

OXYHEMOGLOBIN

A salt or ester of oxaloacetic acid. Oxaloacetate is a four-carbon molecule found in the mitochondrion that condenses with acetyl coenzyme A, an important molecule in metabolism, to form citrate in the first reaction of the citric acid cycle (Krebs cycle). Oxaloacetate is an intermediate in several important metabolic pathways, including the citric acid cycle, the glyoxylate cycle, the urea cycle, gluconeogenesis, and amino acid metabolism.

. . . .

The interaction between oxygen molecules and the various substances they come in contact with. More precisely, oxidation can be defined as the loss of electrons when two or more substances interact, or an increase in oxidation state by a molecule, atom, or ion. Examples of oxidation include the spoiling of fresh fruit and the rusting of an automobile.

. . . .

A bright-red chemical complex of oxygen and the protein hemoglobin, present in arterial blood, which transports oxygen to the tissues. Oxyhemoglobin is produced when oxygen binds to the heme component of hemoglobin in red blood cells during a process called physiological respiration. This process occurs in the pulmonary capillaries that surround the alveoli (air sacs) of the lungs. Oxyhemoglobin is also known as *oxygenated hemoglobin*.

PALISADE MESOPHYLL

. . . .

PARASITISM

. . . .

PARATHYROID GLAND

A tissue system of the chlorenchyma in well-differentiated broad leaves composed of closely spaced, elongated palisade cells arranged in radial columns. It is in these cells that much of the photosynthesis in a tree takes place, as the cylindrical shape of palisade cells allows a large amount of light to be absorbed by chloroplasts, which are concentrated in the palisade mesophyll. In mesophytes (terrestrial plants that adapt to neither a dry nor a wet environment), the palisade mesophyll is found together with spongy mesophyll and is usually on the upper (adaxial) side of the leaf. In xerophytes (plants that survive in an environment that lacks water), it is found on both sides of the leaf and often forms the bulk of the mesophyll.

. . . .

A symbiotic relationship between organisms of different species where one organism (the parasite) benefits at the expense of another organism usually of different species (the host). The close association may also lead to the injury of the host.

. . . .

Any of the four small endocrine glands located in the neck that produce parathyroid hormone, which assists in regulating calcium ion level in blood necessary for normal functioning of neurons. The size of a grain of rice, each gland is 3 to 4 millimeters in diameter and weighs approximately 30 milligrams.

PARENCHYMA

. . . .

PARTHENOGENESIS

. . . .

PARTIAL PRESSURE

The functional tissue or cells of an organ or a gland in the body. This is in contrast to the stroma, which is the structural tissue of organs, specifically, the supporting or connective tissues. In plants, the parenchyma is tissue composed of thin-walled, living cells. Parenchyma is the most common and versatile ground tissue, and forms the mesophyll (internal layers) of leaves, as well as the cortex (outer layers) and pith (innermost layers) of stems and roots, the pulp of fruits, and the endosperm of seeds.

. . . .

Originally from the Greek for "virgin birth"; it is an asexual form of reproduction in which embryonic growth and development occur without fertilization. Many plant species and a few vertebrate and invertebrate animal species can undergo parthenogenesis.

. . . .

The individual pressure exerted independently by a particular gas within a mixture of ideal gases. It is the pressure that a gas would have if it alone occupied the volume. The sum of the partial pressures of the individual gases in the mixture is the total pressure of a gas mixture. Therefore, a gas's partial pressure equals the total pressure multiplied by the fractional composition of the gas in the mixture.

PASSIVE TRANSPORT

. . . .

PATHOGEN

. . . .

PEDUNCLE

The movement of small molecules across membranes of a cell. Expenditure of energy is not required. Passive transport is dependent on the permeability of the cell membrane. The four main kinds of passive transport are:

1. Diffusion is the net movement of molecules from an area of high concentration to an area with lower concentration.
2. Facilitated diffusion is the movement of molecules across a cell membrane with the assistance of special transport proteins that are embedded within the cellular membrane.
3. Filtration is the movement of water and solute molecules across a cell membrane as a result of hydrostatic pressure generated by the cardiovascular system as blood is pumped through the body's blood vessels.
4. Osmosis is the diffusion of water molecules across a semipermeable membrane.

. . . .

An infectious agent or a germ (especially microorganisms such as viruses, bacteria, prions, or fungi) that causes disease or illness to its host. The host can be an animal, a plant, or another microorganism.

. . . .

The stalklike or stemlike connecting part by which an abnormal growth of tissue (a polyp or tumor) is attached to normal tissue.

PELAGIC

....

PELAGIC ZONE

....

PERCENT VIABILITY

Living or growing in or on oceanic waters, neither close to the bottom nor near to the shore. The word *pelagic* comes from the Greek meaning "open sea."

. . . .

An imaginary cylinder or water column that goes from the surface of the sea almost to the bottom. The pelagic zone of the ocean begins at the low tide mark and includes the entire oceanic water column. The zone occupies 320 million cubic miles (1,330 million cubic kilometers). It has a mean depth of 2.29 miles (3.68 kilometers) and a maximum depth of 6.8 miles (11 kilometers).

. . . .

The percentage of viable cells. The calculation for percent viability = (number of viable cells counted ÷ total number of cells counted) × 100.

PERISTALSIS

· ·

PERMAFROST

· ·

A series of organized muscle contractions that moves food through the digestive system. Peristalsis begins in the esophagus when food is swallowed. Strong, wavelike motions of the smooth muscle in the esophagus move the food to the stomach, where it is churned into chyme, a liquid mixture. The chyme is then mixed back and forth in the small intestine, allowing nutrients to be absorbed into the bloodstream through the walls of the small intestine. The process concludes in the large intestine where water from the undigested food material is absorbed into the blood-stream. The remaining waste products are excreted from the body through the rectum and anus. Peristalsis also transports urine from the kidneys into the bladder, and bile from the gallbladder into the duodenum.

. . . .

Permanently frozen soil, sediment, or rock that is at or below the freezing point of water 0°C (32°F) for two or more years. Its classification is not based on moisture or ground cover, but on temperature only. Most permafrost is located in high latitudes, in land close to the north and south poles. Permafrost accounts for 0.022% of total water, comprises 24% of the land in the northern hemisphere, and stores massive amounts of carbon.

. . . .

pH

....

PHAEOPHYTE

....

PHAGOCYTE

P

The negative logarithm of the hydrogen ion concentration [H+]. The pH scale, which ranges from 0 to 14, measures the degree to which a solution is acidic or alkaline. Solutions with a pH less than 7 are acidic, and solutions with a pH greater than 7 are alkaline. Pure water, which has a pH of 7, is neutral. Each whole pH value below or above 7 represents a 10-fold change in the acidity or baseness of the solution (e.g., a solution with a pH of 5 is 10 times more acidic than a pH of 6).

• •

Any of the multicellular algae of the phylum *Phaeophyta*, which contain a brown pigment. Almost all phaeophytes are marine. Though they have traditionally been classified as plants, phaeophytes are not closely related to land plants, as their cells contain different pigments, such as chlorophyll *c* and fucoxanthin.

• •

A cell, such as a white blood cell, that protects the body by ingesting and absorbing waste material, harmful microorganisms, dead or dying cells, or other foreign bodies in the bloodstream and tissues. Therefore, phagocytes are crucial in fighting infections and in maintaining immunity. The name comes from the Greek *phagein*, meaning "to eat" or "to devour." Phagocytes are highly developed within vertebrates; approximately six billion phagocytes are present within one liter of human blood.

PHAGOCYTOSIS

. . . .

PHENOTYPE

. . . .

PHEROMONE

The action of phagocytes in ingesting and destroying cells; the cellular process of engulfing solid particles (bacteria, dead tissue cells, or other foreign bodies) by the cell membrane to remove pathogens and cell debris.

· ·

The composite of an organism's observable physical and/or biochemical characteristics or traits such as height, eye color, and blood type. A phenotype is anything that is part of a living organism's discernible structure, function, or behavior. The genotype is the genetic contribution to the phenotype. Phenotypes are largely determined by the genotype, the influence of environmental factors, and the interactions between the two.

· ·

A chemical substance that, when secreted or excreted into the environment, serves as a stimulus, triggering one or more social or behavioral responses, often from members of the same species. There are many types of pheromones that affect behavior or physiology, including alarm pheromones, food trail pheromones, and sex pheromones. Pheromones relate only to multicellular organisms; some vertebrates and plants communicate by using pheromones, and their use among insects has been well documented. Pheromones are also known as *ectohormones*.

PHLOEM

. . . .

PHOSPHORYLASE KINASE

. . . .

PHOTOAUTOTROPH

P

A tissue in a vascular plant that transports organic nutrients—particularly sucrose, a sugar—from the photosynthetic organ (e.g., the leaves) to all the parts of the plant where needed. The phloem is primarily concerned with translocation, which is the transport of soluble organic material produced during photosynthesis.

· ·

An enzyme (serine/threonine-specific protein kinase) that catalyzes the production of glycogen phosphorylase to release glucose-1-phosphate from glycogen.

· ·

An organism, typically a plant, that uses light as an energy source to convert inorganic materials into organic materials to carry out various cellular metabolic functions such as food synthesis, biosynthesis, and respiration. Photoautotrophs carry out photosynthesis in order to capture light as an energy source. They convert energy from sunlight, carbon dioxide, and water into organic materials. Green plants and photosynthetic bacteria are examples of photoautotrophs. Photoautotrophic organisms are sometimes referred to as holophytic.

PHOTOHETEROTROPH

. . . .

PHOTORECEPTOR

. . . .

PHOTORESPIRATION

An organism that uses light for most of its energy. To satisfy their carbon requirements, photoheterotrophs use organic compounds from the environment since they cannot use carbon dioxide as their sole carbon source. Other compounds, such as carbohydrates, fatty acids, and alcohols, are used by photoheterotrophs for organic food. Purple nonsulfur bacteria, green nonsulfur bacteria, and heliobacteria are examples of photoheterotrophs.

· ·

A nerve ending, cell, or group of cells specialized to sense or receive light. A photoreceptor cell is a type of neuron that converts light into signals that can stimulate biological processes. These cells are found in the retina of the eye. Rods and cones are two classic photoreceptor cells; each contributes information that enables organisms to process visual detail and to form a representation of the visual world, thus enabling sight.

· ·

A process that occurs when the amount of carbon dioxide entering a plant is reduced, and the amount of oxygen produced by photosynthesis increases. Photorespiration takes place on days of high light intensity, dryness, and heat when a plant is forced to close its stomata to prevent excess water loss. Photorespiration thus attempts to produce carbon dioxide and acts as a check on photosynthesis and on the productivity of the plant. It consumes chemical energy rather than produces it.

PHOTOTROPH

· ·

PHYLOGENY

· ·

PHYTOCHROME

An organism (usually a plant) that performs photon capture to acquire energy. Sunlight is used as its principal source of energy in order to carry out various cellular metabolic processes. The most recognized phototrophs are autotrophs, also known as photoautotrophs.

. . . .

The study of evolutionary relationships among species and populations as they change through time, especially in reference to lines of descent and relationships among broad groups of organisms. Most phylogenies, however, are classified as hypotheses based on indirect evidence. Nevertheless, it is universally accepted in the scientific community that true phylogenies are discoverable at least in theory.

. . . .

A cytoplasmic pigment of green plants that is used to detect light. It absorbs light and regulates responses, including dormancy, time of flowering based on the length of day and night, seed germination, elongation of seedlings, and the production of chlorophyll, as well as the size, shape, and number of leaves. Phytochrome is found in the leaves of most plants.

PHYTOPLANKTON

· ·

PINOCYTOSIS

· ·

Microscopic organisms that live in both saltwater and freshwater environments. Some phytoplankton are bacteria whereas others are protists, but most are single-celled plants. Due to the presence of chlorophyll within their cells, phytoplankton may appear as a green discoloration of the water when present in high numbers. Like land plants, phytoplankton obtain energy through photosynthesis; therefore, they must live in the upper sunlit layer of a body of water. Accountable for half of all photosynthetic activity on Earth, phytoplankton are responsible for a large amount of the oxygen present in the atmosphere—approximately half of the total amount produced by all plant life.

· ·

A process in which cells ingest extracellular fluid and its contents by forming narrow channels through the cell membrane, which close and break off to form fluid-filled vacuoles in the cytoplasm, and subsequently fuse with lysosomes to hydrolyze (break down) the contents. This process requires a lot of energy in the form of adenosine triphosphate (ATP). Used principally for the absorption of extracellular fluids, pinocytosis produces very small vesicles, compared with phagocytosis. Pinocytosis is nonspecific in the substances that it transports, as the cell takes in all surrounding fluids that are present.

· ·

PLASMA MEMBRANE

....

PLASMODESMA

....

A cell's outer membrane that surrounds the cytoplasm and separates the interior of the cell from the outside environment. It is made up of two layers of phospholipids with embedded proteins. Selectively permeable, the plasma membrane is able to regulate the movement of substances into and out of the cell, thus facilitating the transport of materials needed for survival. The movement of substances through the membrane can occur through passive transport, which occurs without the input of cellular energy, or through active transport, which requires the cell to expend energy during transportation. Many molecules cross the plasma membrane by diffusion and osmosis.

. . . .

A microscopic cytoplasmic canal that passes through plant cell walls, enabling transport and direct communication of molecules between adjacent plant cells. There are two forms of plasmodesmata (plural of plasmodesma). Primary plasmodesmata are formed during cell division, when portions of the endoplasmic reticulum become trapped in the new wall that divides the parent cell. Secondary plasmodesmata are formed between existing cell walls of mature, nondividing cells.

. . . .

PLASMODIUM

. . . .

PLASMOLYSIS

. . . .

PLATELET

P

The parasite responsible for malaria. A type of protozoa, plasmodium is a single-celled organism that is able to divide only within a host cell. The parasite has two hosts in its life cycle: a vector, which is typically a mosquito, and a vertebrate host. Currently, over 200 species of this genus are recognized; at least 11 of these species infect humans, while other species infect other animals, including monkeys, rodents, birds, and reptiles.

. . . .

The process in plant cells where contraction or shrinkage of the protoplasm away from the cell wall of a living plant or bacterium is caused by the loss of water through osmosis. This results in gaps between the cell wall and cell membrane.

. . . .

Any of the disk-shaped, clear cell fragments that circulate in the blood of mammals. Platelets are produced from large bone marrow cells called megakaryocytes. The principal function of platelets is to prevent bleeding. Because of their diminutive size (2–3 µm in diameter) platelets make up just a tiny fraction of the blood volume. The average lifespan of a platelet is approximately five to nine days. Normal platelet count is 150,000 to 350,000 per microliter of blood. If platelet count is too low, excessive bleeding can occur; if platelet count is too high, blood clots can form, which may obstruct blood vessels and result in harmful events such as a stroke, myocardial infarction, or pulmonary embolism. Platelets are also known as *thrombocytes*.

PLEIOTROPY

....

POLARITY

....

Occurs when a single gene controls or influences multiple phenotypic traits. A mutation in a pleiotropic gene may have an effect on some or all traits concurrently, and can become a problem when selection on one trait favors one allele (specific version of the gene), while the selection on another trait favors another allele. A common example of pleiotropy is the human disease phenylketonuria (PKU), which can be caused by any of a large number of mutations in a single gene.

· ·

The asymmetric organization of most of the physical aspects of the cell, including the cytoskeleton, intracellular organelles, and cell surface. Differentiation in the ways these parts are asymmetrically organized helps to determine cell function and cell type. Nearly all cell types exhibit some form of polarity, which enables them to perform specialized functions like migration, axis formation, and asymmetric cell division. Many of the known molecules involved in cell polarity are conserved across animals. Typical examples of polarized cells include epithelial cells with apical-basal polarity, neurons in which signals propagate in one direction from dendrites to axons, and migrating cells.

· ·

POLYANDRY

P

The practice of one female having more than one male mate. Polyandry, which means "many males," is a rare mating system that is often accompanied by a reversal of sexual roles in which females compete for mates and males perform parental duties. Two types of polyandry have been documented. In simultaneous polyandry, a female occupies a large territory that contains the smaller nesting territories of two or more males, which she may help in defending. In sequential polyandry, the most typical form of this mating system, a female mates with a male and then terminates the relationship, leaving him to tend to the young while she repeats this process with another male. Polyandry can be found in field crickets, frogs, and some mustelids (weasel family). It also occurs in some primates such as marmosets, marsupials such as bandicoots, insects such as honeybees, fish such as pipefish, and approximately 1% of all bird species (e.g., jacanas).

. . . .

POLYGENIC TRAIT

. . . .

POLYGYNY

. . . .

Any of the traits that are determined by two or more genes at different loci on different chromosomes. These genes are called polygenes. Examples of polygenic traits are human skin, hair, and eye color, because they are influenced by more than one allele at different loci. Many genes factor into determining a person's natural skin, hair, and eye color, so modifying only one of those genes changes the color only slightly, resulting in continuous gradation in the expression of these traits.

• •

The practice of one male having more than one female mate at a time. Polygyny is considered the fundamental mating system of animals. A variety of methods for practicing polygyny exists. In female defense polygyny, as seen in marine amphipods, males compete among each other for the right to mate with a cluster of females. In resource defense polygyny, seen in African cichlid fish, males compete among each other for valuable resource territories that females need. A third method is scramble competition polygamy, seen in orangutans, where females are widely dispersed or fertility is time-limited. In this scenario, males attempt to outdo each other in order to gain potential mates. Mating success is reserved for males that are the most determined, resilient, and perceptive, rather than the most aggressive and physically imposing.

• •

**POLYMERASE CHAIN REACTION
(PCR)**

A biochemical technology in molecular biology used to reproduce selected sections of DNA or RNA across several orders of magnitude, generating thousands to millions of copies of a particular DNA sequence. PCR is a common technique used in medical and biological research labs for a variety of applications, such as DNA cloning for sequencing, diagnosis of hereditary diseases, identification of genetic fingerprints, and detection and diagnosis of infectious diseases.

Three major steps, which are repeated for 30 or 40 cycles, are involved in a PCR. Using an automated cycler, a device that rapidly heats and cools the test tubes containing the reaction mixture, each step takes place at a different temperature:

1. Denaturation: At 201.2°F (94°C), the double-stranded DNA melts and disrupts the hydrogen bonds between complementary bases, yielding two pieces of single-stranded DNA molecules.
2. Annealing: At 129.2°F (54°C), the short single-stranded DNA sequences that are synthesized to correspond to the beginning and ending of the DNA stretch to be copied (the primers) pair up with the sequence of DNA to be copied (the single-stranded template). Stable DNA–DNA hydrogen bonds are formed only when the primer sequence very closely matches the template sequence. On the small length of double-stranded DNA (the joined primer and template), the polymerase (enzyme that moves along the segment of DNA) binds to the primer-template hybrid and starts copying the template.
3. Extension: At 161.6°F (72°C), the DNA polymerase synthesizes a new DNA strand complementary to the DNA template strand; this is coupled to the primer, making a double-stranded DNA molecule.

POLYMORPHISM

. . . .

POLYP

. . . .

The occurrence of more than one form or morph when two or more different phenotypes exist in the same population of a species. To be classified as polymorphic, morphs must occupy the same habitat at the same time and belong to a random mating population where there are no mating restrictions and where all individuals are potential partners. Common in nature, polymorphism is associated with biodiversity, genetic variation, and adaptation. Polymorphism results from evolutionary processes and is modified by natural selection. However, it is also heritable; in genetic polymorphism, the genetic makeup determines the morph.

. . . .

A small abnormal growth of tissue, typically benign, projecting from a mucous membrane. If a narrow elongated stalk is attached to the polyp, the polyp is pedunculated. If no stalk is present, the polyp is sessile. Polyps are commonly found in parts of the body where mucous membranes exist, such as the colon, stomach, small intestine, nose, sinuses, urinary bladder, uterus, and cervix.

. . . .

POLYPEPTIDE

. . . .

POLYPLOIDY

. . . .

A molecular chain of amino acids. The amino acids are linked covalently by peptide bonds to form polypeptides. Proteins consist of one or more polypeptide molecules. A free amino group is located on one end of every polypeptide. This is called the amino terminal or N-terminal. This is the end that amino acids are added onto when mRNA (messenger RNA) is translated in protein synthesis. The other end, which has a free carboxyl group, is called the carboxyl terminal or C-terminal.

. . . .

A cell or organism that possesses more than two complete sets of homologous chromosomes. Polyploidy occurs from abnormal cell division, most often during metaphase I in meiosis or during mitosis. Polyploidy is generally rare in mammals but is common among certain groups of insects, fish, and amphibians. Many cancer cells are polyploids. Polyploidy is also known as whole genome duplication. Polyploid types are labeled according to the number of chromosome sets in the nucleus:

- triploid (three sets)
- tetraploid (four sets)
- pentaploid (five sets)
- hexaploid (six sets)
- octaploid (eight sets)
- decaploid (10 sets)
- dodecaploid (12 sets)

. . . .

POLYRIBOSOME

. . . .

A cluster of ribosomes, connected by a strand of mRNA, that function as a unit in protein synthesis. Polyribosomes read one strand of mRNA simultaneously, helping to synthesize the same protein at different spots on the mRNA. Polyribosomes may appear as clusters, as linear polysomes, or as circular rosettes in microscopes, but emerge as circular in live isolated cells because mRNA is able to form circular structures, which create a cycle of rapid ribosome recycling and utilization of ribosomes. The three forms of polyribosomes are free, cytoskeletal-bound, and membrane-bound. Polyribosomes are also known as *polysomes* or *ergosomes*.

• •

POLYSACCHARIDE

A long carbohydrate molecule formed by repeating units linked together by glycosidic bonds. Polysaccharides may have a molecular structure that is either branched (forms soluble in water and that make pastes) or linear (compounds that are often packed together to form a rigid structure). Polysaccharides composed of many molecules of one sugar or one sugar derivative (monosaccharides) are called homopolysaccharides or homoglycans. Examples include glycogen and starch (the storage carbohydrates of animals and plants, respectively) and cellulose (the structural component of most plants). Polysaccharides composed of more than one type of monosaccharide are called heteropolysaccharides or heteroglycans. Examples include pectin, lignin, glycoproteins, glycolipids, and mucopolysaccharides. Polysaccharides are also known as glycans. Polysaccharides are characterized by the following chemical properties:

- not sweet in taste
- insoluble in water
- do not form crystals when desiccated
- compact and not osmotically active inside the cells
- can be extracted to form white powder
- general chemical formula of $C_x(H_2O)_y$, where x is usually a large number between 200 and 2,500

. . . .

POLYUNSATURATED FATTY ACID

. . . .

PONS

. . . .

An unsaturated fatty acid that contains more than one double or triple valence bond per molecule in its hydrocarbon chain. Polyunsaturated fatty acids (PUFAs) are found primarily in fish, seeds, bananas, nuts, and vegetable oils. Because of their chemical structure, PUFAs have a lower melting point than monounsaturated fatty acids (fatty acids that have one double bond in the fatty acid chain). They contain essential fatty acids like omega-3 and omega-6 acids, which the body needs but cannot produce. PUFAs are critical structural components of cell membranes and nerve tissue. They also help regulate the production of prostaglandin, a chemical that helps the body's inflammatory functions.

A structure located on the brain stem. From the Latin meaning "bridge," the pons, measuring approximately 2.5 cm in length, is a portion of the hindbrain that connects the cerebral cortex with the medulla oblongata. As a communications and coordination center, the pons relays messages between the two hemispheres of the brain. The pons contains nuclei involved in key functions of the body, including sleep, respiration, swallowing, bladder control, hearing, equilibrium, taste, eye movement, facial expressions, facial sensation, and posture.

POSITIVE CONTROL

....

POSTZYGOTIC BARRIER

....

A representative sample or treatment for which the result is known with certainty. Positive controls are often used to assess test validity. They give credibility to an experiment by using an experimental treatment that has already been recognized in producing a specific effect, and provide an overall reliability measure of the experiment's data. If the expected result is not produce by the positive control, this indicates that something may have gone wrong with the experimental procedure, and the experiment is repeated.

. . . .

Reproductive barrier that forms when interspecies mating occurs and hybrid zygotes are formed. In the postzygotic barrier of hybrid inviability, the hybrid offspring of two species are frequently abnormal, and most do not survive birth or germination. For example, in crosses between different species of irises, the embryos die before seeds form. In the postzygotic barrier of hybrid sterility, those species that do survive often never reach sexual maturity and are usually unable to reproduce. Mules are sterile hybrids formed by mating a female horse with a male donkey. If mating between two hybrids creates a second hybrid generation, this generation may be unable to reproduce because of hybrid breakdown. This postzygotic barrier has been seen in sunflower hybrids.

. . . .

POTENTIAL ENERGY

....

PRESSURE POTENTIAL

....

The energy stored in an object that is derived from position rather than motion. An object may have the potential for doing work as a result of its position. Potential energy exists when a force (called a restoring force) acts upon an object, restoring it to a lower-energy configuration. For example, when a metal spring is stretched to the left, it exerts a force to the right so as to return to its original, unstretched position. The energy it took to stretch the spring is stored in the metal as potential energy. If the spring is released, this potential energy will be changed into kinetic energy (the energy of motion) by the restoring force, which is elasticity. The unit for measuring work and energy is the joule (abbreviation: J).

. . . .

A component of the total water potential due to the hydrostatic pressure that is exerted on water in plant cells. In turgid plant cells, pressure potential usually has a positive value, as the entry of water through the cell wall and cell membrane causes the protoplast to push against the cell wall. By creating this pressure, the plant can maintain turgor, which allows the plant to keep its rigidity. In xylem cells there is a negative pressure potential, or tension, as a result of transpiration (evaporation of water into the atmosphere from leaves and stems). Water at atmospheric pressure has a pressure potential of zero.

. . . .

PREZYGOTIC BARRIER

The prevention of mating or fertilization between members of different species. A prezygotic barrier in which the two species reproduce at different times of the day, season, or year is temporal isolation. Habitat isolation is a prezygotic barrier in which two species whose ranges overlap live in different habitats. As a result, potential mates from the two species do not encounter one another. In behavioral isolation, each group possesses its own characteristic courtship behaviors; therefore reproduction between similar species is prevented. For instance, wood and leopard frogs have vocalizations that attract only females of their species. Unsuccessful mating due to genital organs of different species being incompatible is mechanical isolation. In plants, mechanical isolation often occurs in flowering plants pollinated by insects. Similar to mechanical isolation is gametic isolation, in which the egg and sperm of different species are incompatible, increasing the likelihood that fertilization will not occur.

. . . .

PRIMARY CELL WALL

. . . .

PROCAMBIUM

. . . .

A thin, flexible, and extensible layer formed while the cell is growing. The primary cell wall of most plant cells is semipermeable. Primary walls are composed primarily of polysaccharides (cellulose, pectin, and hemicelluloses). They also contain proteins (expansins) that regulate wall expansion. Functions of the primary cell wall include:

- structural and mechanical support
- maintaining and determining cell shape
- resisting internal turgor pressure of cell
- controlling rate and direction of growth
- regulating diffusion of material

. . . .

Long, narrow cells of a meristem, which consists of undifferentiated tissue from which new cells are formed, as at the tip of a stem or root. The procambium is located inside the protoderm (a thin outer layer of the meristem, which gives rise to the epidermis) and develops into the primary vascular tissues of xylem and phloem. The procambium also produces a secondary meristem called the vascular cambium.

. . . .

PROGESTERONE

P

A female steroid hormone. Progesterone belongs to a class of hormones called progestogens, and is produced in the ovaries, in the adrenal glands, and in the placenta during pregnancy. It is also stored in adipose (fat) tissue. One of progesterone's most important functions is to help prepare the endometrium (lining of the uterus) to receive and nourish an implanted fertilized egg during the second half of the menstrual cycle. If the egg is not fertilized, estrogen and progesterone levels drop, the endometrium breaks down and menstrual bleeding begins. If a pregnancy occurs, high levels of progesterone are produced in the placenta, starting near the end of the first trimester and continuing until the baby is born. Progesterone levels are approximately 10 times higher in pregnant women than in women who are not pregnant. The combination of high estrogen and progesterone levels suppresses further ovulation during pregnancy. During pregnancy, progesterone also encourages the growth of milk-producing glands in the breasts.

· ·

PROKARYOTE

. . . .

PROKARYOTIC CELL

. . . .

Single-celled organisms without a cell nucleus, mitochondria, or any other membrane-bound organelles. Prokaryotes are the earliest and most primitive forms of life on Earth. Belonging to two taxonomic domains, prokaryotes include bacteria and archaeans. Prokaryotes have been found in all types of environments, including extreme habitats such as hydrothermal vents, hot springs, swamps, wetlands, and the guts of animals.

. . . .

A cell that lacks a membrane-bound nucleus. Prokaryotic cells are either spherical, spiral, or rod-shaped. Prokaryotic cells are not as complex as eukaryotic cells, and the distinction between the two is that eukaryotic cells have a true nucleus containing their DNA, whereas prokaryotic cells do not have a membrane-bound nucleus. This is considered one of the most important distinctions among organisms. Instead of being divided into different cellular compartments, all the intracellular components (proteins, DNA, and metabolites) of prokaryotes are coiled up in a circular loop called the nucleoid, which is a region of the cytoplasm. Unlike eukaryotes, which go through intricate replication processes, prokaryotic cells divide by binary fission.

. . . .

PROMOTER

. . . .

PROPHASE

. . . .

P

A sequence of DNA needed to turn a gene on or off. The process of transcription, creating a complementary RNA copy of a sequence of DNA, is initiated at the promoter. For the transcription to take place, RNA polymerase (the enzyme that synthesizes RNA) must attach to the DNA near a gene. Usually found near the beginning of a gene, promoters provide a secure initial binding site for RNA polymerase, which is used to make a messenger RNA (mRNA) molecule. In prokaryotes, the promoter is recognized by RNA polymerase; however, in eukaryotes, the process is more complicated, and a number of different factors are necessary for the binding of an RNA polymerase II to the promoter.

. . . .

A stage of mitosis in which strands of chromatin (genetic material composed of DNA and proteins) form into discrete chromosomes. During prophase, the nuclear envelope breaks down and mitotic spindles migrate to opposite poles of the cell. Compared with interphase, many consider prophase to be the first true step of the mitotic process.

. . . .

PROTIST

....

PROTODERM

....

P

A diverse group of predominantly unicellular eukaryotic microorganisms. Their simple cellular organization distinguishes protists from other eukaryotes, such as fungi, animals, and plants. Protists are grouped into three major categories: protozoa, algae, and lower fungi. Protists need a water-based environment in which to live. It can be a variety of habitats such as saltwater or freshwater, damp soil, snow, or even polar bear hairs. All contain mitochondria for cellular respiration and are therefore aerobic. Some have chloroplasts and can perform photosynthesis. Most protists reproduce or grow by mitosis, whereas others reproduce by meiosis and fertilization.

. . . .

The outermost layer of cells across the top of the apical meristem and leaf primordium. The protoderm is a primary meristem in vascular plants that gives rise to the epidermis and is also associated with subepidermal tissues. Protoderms are also known as *dermatogens*.

. . . .

PROTON-MOTIVE FORCE

. . . .

PROTOPLAST

. . . .

PROTOSTOME

The storing of energy as a combination of proton and voltage gradients across a membrane. The proton-motive force is generated by an electron transport chain that acts as a proton pump to pump protons (hydrogen ions) out across the membrane, using the energy in electrons from an electron carrier. By pumping protons across the inner membrane, a difference in concentration is created where positively charged hydrogen ions are more concentrated outside the membrane than they are inside, and the ions cannot diffuse back across the membrane. This concentration gradient is called the proton-motive force.

• •

The living material of a plant, bacterial, or fungal cell, including the protoplasm and plasma membrane, whose cell wall has been completely or partially removed using either mechanical or enzymatic means.

• •

Any of a major group of animals defined by its embryonic development, in which the mouth opening is the first to be formed, followed later by the anus. The word *protostome* comes from the Greek for "first mouth." Protostomes (such as mollusks, annelids, and arthropods) are bilaterally symmetrical. In contrast to protostomes, deuterostomes (such as echinoderms and chordates) develop the anus first, followed by the mouth.

PSEUDOCOELOM

· ·

PSEUDOCOELOMATE

· ·

A closed, fluid-filled main body cavity in proto-somes. The pseudocoelom is located between the digestive tract and the body, allowing the digestive system and the body wall to move independently. Surrounding a protosome's internal organs, the pseudocoelom assists with distribution of nutrients and removal of waste. Supporting the body, it acts as a hydrostatic skeleton to hold the major body organs, circu-late nutrients, and maintain body shape.

. . . .

Any invertebrate animal with a three-layered body and a pseudocoelon, which is a fully functional body cavity. Tissue derived from the mesoderm only partly lines the fluid-filled body cavity of these animals. Thus, organs are held in place loosely. All pseudocoelomates are protostomes; however, not all protostomes are pseudocoelomates. An example of a pseu-docoelomate is the roundworm. Important characteristics of pseudocoelomates are:

- a complete digestive tract
- a body wall of epidermis and muscle
- lack of a vascular blood system
- lack of a skeleton
- lack of circulatory and respiratory organs

. . . .

PSEUDOPODIUM

. . . .

PUNCTUATED EQUILIBRIUM

. . . .

Temporary projection or retractile process of the cytoplasm of a cell that functions as an organ of locomotion or in taking up food. Pseudopodia are formed by amoebas and cells of higher animals like white blood cells. During feeding, pseudopodia trap prey by either capturing it in a sticky net or flowing over a surface and engulfing prey. There are four types of pseudopodia: lobopodia, characteristic of amoebas, are segmented and wormlike, and have legs with hooked claws; filopodia are slender and occasionally form simple, branched networks; reticulopodia are branching filaments that fuse to form netlike structures; and axopodia are long and sticky, composed of an outer layer of flowing cytoplasm that surrounds a stiff, internal rod containing a bundle of microtubules.

· ·

A theory that proposes that instead of species exhibiting little net evolutionary change for millions of years, this virtual standstill or equilibrium tends to be punctuated by rare and geologically rapid bursts of change that result in new species that leave few fossils behind. According to this idea, the changes leading to a new species are more likely to occur at the edge of a population, where, geographically, a small group can easily become separated from the mainstream population and undergo transformations that can produce a new, noninterbreeding species. Contrasting with the theory of punctuated equilibrium is phyletic gradualism, which theorizes that evolution is slow, uniform, and gradual.

· ·

PUNNETT SQUARE

A simple, graphical diagram that is used to calculate the mathematical probability of inheriting a specific trait. Named after geneticist Reginald C. Punnett, the Punnett square is a tabular summary of all the potential combinations of genotypes that can occur in offspring, given the genotypes of their parents. It also depicts the correct odds for the genotype outcomes of each of the offspring. The Punnett square is a visual representation of Mendelian inheritance, the scientific theory of how hereditary characteristics are passed from parent organisms to their offspring.

· ·

QUATERNARY STRUCTURE

. . . .

The structure formed by the combination of two or more polypeptide chains that form subunits. The arrangement of the subunits gives rise to a stable structure. Stabilizing forces of the quaternary structure include interactions between side chains of the subunits, such as hydrophobic interaction among nonpolar side chains and electrostatic interactions between ionic groups of opposite charge. An example of a protein with quaternary structure is hemoglobin, a group of four globulin protein molecules that are bound by the iron in heme molecules.

. . . .

RADICAL SYMMETRY

. . . .

A basic body plan in which an organism can be divided into similar halves. An organism displays radial symmetry when it has a regular arrangement of body parts around a central axis; therefore, the organism exhibits no left or right sides, only a top and a bottom (dorsal and ventral surfaces). From anywhere on the organism, the organism can be cut from one side through the center to the other side and this cut would produce two equal halves. Animals such as sea anemones, jellyfish, and sea stars exhibit radial symmetry.

. . . .

RADIATION

A process in which energetic particles or waves travel outward in all directions through a medium or space. Radiation can be naturally occurring or man-made. While there are various types of radiation, two types are generally differentiated in the way they interact with normal chemical matter:

1. Nonionizing radiation has energy to excite atoms and to make them move more rapidly, but not enough to cause ionization (physically altering the atoms). Examples include microwaves, radio waves, and visible light.

2. Ionizing radiation is the most energetic form of radiation; it can remove tightly bound electrons from the orbit of an atom, causing the atom to become charged or ionized and damaging the DNA within living cells. Examples include X-rays, gamma rays, and alpha and beta particles.

. . . .

REACTANT

. . . .

REACTION CENTER

. . . .

A substance that enters into and is altered in the course of a chemical reaction. A limiting reactant is the substance that is totally consumed when a chemical reaction is complete. The reaction will stop when all of the limiting reactant is consumed. It limits the amount of product that can be formed since the reaction cannot proceed further without it. In a chemical reaction, the excess reactant is the reactant that remains when a reaction stops when the limiting reactant is completely consumed. Since there is nothing with which it can react, the excess reactant remains. Reactants are also known as *reagents*.

. . . .

A complex system consisting of proteins, pigments, and cofactors assembled together to facilitate light energy and electron transfer in plants. Reaction centers execute the primary energy conversion reactions of photosynthesis and are present in all green plants, algae, and many bacteria.

. . . .

RECOMBINANT DNA

. . . .

RECOMBINATION

DNA sequences that result from the use of laboratory methods to combine two or more different strands of DNA from multiple sources in order to create a new strand of DNA that would not otherwise be found in biological organisms. The DNA sequences used in the construction of recombinant DNA molecules can arise from any species; for example, fungal DNA and human DNA can be combined. Recombinant DNA can be created from three different methods: transformation, nonbacterial, and phage introduction. All of these methods have a similar technique, and their goal is the same: to introduce recombinant genes into a host cell along with an expression factor so that the host cell expresses the desired protein. Recombinant DNA is also known as *chimeric DNA*.

. . . .

The crossing-over of chromosomes during meiosis, in which DNA is exchanged between a pair of chromosomes to form new molecules of DNA, encoding a novel set of genetic information. Because of recombination, two genes that were on separate chromosomes and previously unlinked can now become linked. The opposite can happen as well; linked genes may become unlinked. Recombination can take place between similar molecules of DNA (homologous recombination) or dissimilar molecules (nonhomologous end joining). Recombination also serves as a method of DNA repair in prokaryotes (bacteria) and eukaryotes. Recombination is one of the important processes that promote and increase genetic diversity between generations.

REDUCTION POTENTIAL

. . . .

REGULATORY SEQUENCE

A measure of the capacity of a compound to gain electrons and thereby be reduced. The symbol for reduction potential is $E°'$. The higher the $E°'$ value, the stronger the tendency for a compound to gain electrons. In reduction potential, electrons flow from a substance whose half reaction has the lower reduction potential to a substance whose half reaction has the higher reduction potential. Reduction potentials determine the probability that a half reaction will take on the role of the reduction in an oxidation-reduction (redox) reaction. When paired in a redox reaction, the half reaction with the higher $E°'$ will act as the reduction reaction, and the other half reaction with the lower $E°'$ will act as the oxidation reaction. Reduction potential is measured in volts (V) or millivolts (mV). Reduction potential is also known as *redox potential* and *oxidation/reduction potential*.

. . . .

A DNA sequence where regulatory proteins bind to control the rate of transcription (the copying of DNA into messenger RNA in gene expression). In a regulatory sequence, regulatory proteins bind to regulatory regions, which are short stretches of DNA that are positioned in the genome, typically located a short distance upstream from the gene they regulate. Through this binding, the regulatory proteins can recruit RNA polymerase, a protein complex. In this way, they control gene expression and consequently, protein biosynthesis. A regulatory sequence is also known as a regulatory region or a regulatory area.

REPRESSIBLE SYSTEM

. . . .

REPRESSOR

. . . .

RESPONSE ELEMENT

A system in which the synthesis of a specific enzyme can be repressed (switched off) by the high concentration of its own product. In a repressible system, the product is called a corepressor, which is a substance that suppresses gene expression. When the corepressor is united with an aporepressor (inactive repressor), it forms a functional repressor, which prevents gene transcription and inhibits protein synthesis. This blocking is called repression.

. . . .

One of the proteins that bind to specific sites on DNA and prevent transcription of nearby genes. A repressor consists of a DNA-binding protein determined by a regulatory gene, which binds to an operator (a segment of DNA to which a transcription factor protein binds) and blocks the attachment of RNA polymerase to the promoter. This blocking of expression, called repression, inhibits transcription of the genes.

. . . .

Short sequences of DNA within gene promoter or enhancer regions that are capable of binding a specific transcription factor and regulating transcription of genes. Sequences of response elements are required for the functions of the transcription factors.

RESTING POTENTIAL

. . . .

RHIZOID

. . . .

RHIZOME

The electric potential across the membrane of a normal cell at rest before it is stimulated to release the charge. Resting potential is the difference in electric charge between the inside and outside of a neuron's cell membrane. The resting potential for a neuron is between 50 and 100 mV, with a resulting accumulation of excess negative charge inside the cell membrane.

. . . .

A short, thin filament found in plants and fungi that is similar in design and function to the roots of more developed land plants. Rhizoids are a single tissue, whereas roots themselves are organs composed of multiple tissues. Rhizoids anchor the growing (vegetative) body of the organism to a layer of earth beneath the surface soil and are capable of absorbing nutrients.

. . . .

A horizontal underground plant stem that can send out both shoots and roots from its nodes. In a process known as vegetative reproduction, a rhizome is separated into pieces and each piece is able to give rise to a new plant. Farmers and gardeners use this to propagate certain plants, including hops, asparagus, ginger, and irises. Rhizomes are also known as *rootstalks* and *rootstocks*.

RHODOPHYTE

. . . .

Red algae, or rhodophyta. Most rhodophytes are marine, though there are freshwater species. Rhodophytes are multicellular and macroscopic, and can sexually reproduce. The pigment phycoerythrin, which reflects red light and absorbs blue light, is what gives red algae its color. These pigments allow red algae to photosynthesize and live at greater depths than most other algae, because blue light can penetrate water to a greater depth than light of longer wavelengths. However, not all rhodophytes are red; those that have very little phycoerythrin may appear green or blue in color from the chlorophyll and other pigments present in them.

· ·

RIBONUCLEIC ACID (RNA)

A nucleic acid molecule similar to DNA but containing the sugar ribose rather than de-oxyribose. RNA is made up of a long chain of nucleotides, which allow RNA to encode genetic information. Through the process of transcription, the enzyme RNA polymerase synthesizes RNA by using DNA as a template. There are several classes of RNA molecules that play crucial roles in protein synthesis and other cell activities:

- Messenger RNA (mRNA) reflects the exact nucleoside sequence of the genetically active DNA. mRNA carries codes from the DNA in the nucleus to the sites of protein synthesis in the cytoplasm (the ribosomes).
- Transfer RNA (tRNA) is a short-chain type of RNA present in cells. There are 20 varieties of tRNA. Each variety transports amino acids to ribosomes for incorporation into a polypeptide undergoing synthesis.
- Ribosomal RNA (rRNA) is a component of ribosomes that functions as a nonspecific site for making polypeptides.

. . . .

RIBONUCLEOTIDE

. . . .

A nucleotide that contains the sugar ribose and that is found in mRNA. A ribonucleotide is a nucleotide (joined molecules that make up the individual structural units of the nucleic acid RNA) in which a purine or pyrimidine base is linked to a ribose molecule and one phosphate group. In living organisms, the most common bases for ribonucleotides are a group of nitrogen-based molecules that are the molecular building blocks of RNA: adenine (A), which attaches only to uracil (U), and cytosine (C), which attaches only to guanine (G).

. . . .

RIBOSOME

A minute particle, composed of three or four RNA molecules and anywhere from 40 to 80 different proteins, which serves as the site of protein synthesis. Found in all living cells, ribosomes differ in their size, sequence, structure, and ratio of protein to RNA. Ribosomes are very numerous in a cell and account for a large proportion of its total RNA. Ribosomes occur as free particles in prokaryotic and some eukaryotic cells, scattered throughout the cytoplasm. In other eukaryotic cells, they are bound to the endoplasmic reticulum and the nuclear envelope. During protein synthesis, ribosomal molecules of messenger RNA (mRNA) regulate the order of transfer RNA (tRNA) molecules that are bound to amino acids, which ultimately determine the amino acid sequence of a protein. The proteins that are then formed detach themselves from the ribosome site and move to other parts of the cell for use.

. . . .

ROTIFER

Microscopic aquatic animals of the phylum Rotifera. Rotifers are 0.1 to 0.5 mm long, and are common in freshwater environments with a few saltwater species. Rotifers are multicellular animals that exhibit bilateral symmetry. Their body, typically cylindrical, is divided into a head, trunk, and foot. Rotifers have specialized organ systems and a complete digestive tract that includes both a mouth and an anus. The name *rotifer* is derived from the Latin word meaning "wheel bearer"; this makes reference to the crown of cilia around its mouth that, when in rapid motion, resembles a wheel. Rotifers eat particles up to 10 micrometers in size, including dead bacteria, algae, and protozoans. Rotifers reproduce through parthenogenesis, an asexual form of reproduction in which embryonic growth and development occur without fertilization.

. . . .

SALTATORY CONDUCTION

. . . .

SAPROBE

. . . .

The passage of action potentials from one node of Ranvier to the next node of a nerve fiber, rather than along the membrane. The distance between these nodes, or myelin sheath gaps, ranges from 0.2 to 2 mm. The term *saltatory conduction* is from the Latin meaning "to hop or leap." Saltatory conduction is a faster way to travel down an axon than traveling in an axon without myelin. In myelinated axons, action potentials traveling down the axon do not move continuously as waves, but instead jump from node to node, by which process they travel faster than they would otherwise.

· ·

An organism that acts as a decomposer, deriving its nourishment from nonliving organic matter. Saprobes feed on dead and decaying pieces of animals, plants, wood, leaves, litter, and other organic matter. To digest their food, saprobes secrete enzymes that break down organic material and recycle vital nutrients. Saprobes are the most common type of fungi. Many bacteria and protozoa are saprobes as well.

· ·

SARCOMERE

. . . .

The basic unit of a muscle. Sarcomeres are composed of long, fibrous proteins (myofilaments) that slide past each other when muscles contract and relax. Interactions between the thick and thin filaments found in sarcomeres are responsible for muscle contraction. Sarcomeres are the smallest functional units of the muscle fiber, with a resting length of 1.6 to 2.6 μm. Differences in the size, density, and distribution of thick and thin filaments account for a sarcomere's striated appearance. When viewed under polarized light, filaments appear as dark bands (anisotropic) and light bands (isotropic):

- A bands are composed of thick and overlapping thin filaments that are located at the center of a sarcomere; they appear as dark bands.
- I bands are composed of thin filaments that extend from the A band of one sarcomere to the A band of the next sarcomere; they appear as light bands.
- Z lines are borders that separate and link sarcomeres and consist of proteins called connectins, which interconnect thin filaments of adjacent sarcomeres; they appear as a series of dark lines.

SARCOPLASMIC RETICULUM

. . . .

SCLEREID

. . . .

A type of smooth endoplasmic reticulum found in smooth and striated muscle fibers. While the endoplasmic reticulum synthesizes molecules, the function of the sarcoplasmic reticulum is to store and pump calcium ions. Large stores of calcium are sequestered, and then calcium ions are released when the muscle cell is stimulated (during muscle contraction). The sarcoplasmic reticulum absorbs the calcium ions during relaxation.

. . . .

A plant cell with highly thickened, lignified walls. Sclereids are a type of sclerenchyma cell, which are cells with lignified secondary walls that have lost their protoplasm at maturity. They can be grouped into small bundles of sclerenchyma tissue in plants that form durable layers such as seed coats and nut shells. Sclereids are elongated (often branched in shaped) and flexible, with tapered ends. Compared with most fibers, sclereids are relatively short.

. . . .

SENSORY RECEPTOR

Dendrites of sensory neurons that respond to specific stimulus modalities in the internal or external environment of an organism. The sensory receptor functions are the first component in a sensory system. Sensory receptors can be classified by the type of stimulus detected: mechanoreceptors respond to physical force such as touch and pressure, thermoreceptors respond to temperature changes, photoreceptors respond to light, and chemoreceptors respond to dissolved chemicals during taste and smell. They also respond to changes in internal body chemistry such as variations of oxygen, carbon dioxide, or hydrogen ions in the blood. Sensory receptors can also be classified by location:

- Exteroceptors respond to stimuli occurring outside or on the surface of the body. These receptors include those for tactile sensations (touch, pain, and temperature), and those for vision, hearing, smell, and taste.
- Interoceptors respond to stimuli occurring inside the body from internal organs and blood vessels. These receptors are the sensory neurons associated with the autonomic nervous system.
- Proprioceptors respond to stimuli occurring in skeletal muscles, tendons, ligaments, and joints. These receptors collect information concerning body position and the physical conditions of these body parts.

SIGNAL SEQUENCE

. . . .

SIGNAL TRANSDUCTION

. . . .

A short peptide chain (3 to 60 amino acids long) that directs the transport of a protein to certain organelles such as the nucleus, endoplasmic reticulum, and chloroplast. After the proteins are transported, some signal sequences are cleaved from the protein by signal peptidase, which are enzymes that convert membrane proteins and any protein secreted by the cell to their mature form by cutting off their N-terminal signal sequences. Signal sequences are also known as *signal peptides*, *targeting signals*, *transit peptides*, and *localization signals*.

. . . .

The cellular process in which a signal is conveyed to trigger a change in the activity or state of a cell. In signal transduction, a signaling molecule located within the cell activates a specific receptor protein on the cell membrane. A second messenger then transmits the signal into the cell, eliciting a physiological response. The eventual outcome is an alteration in cellular activity and regulation of transcription or of a metabolic process.

. . . .

SIMPLE DIFFUSION

. . . .

SOLUTE

. . . .

A process whereby the movement of small, nonpolar molecules through a permeable membrane is done without the aid of an intermediary such as an integral membrane protein. Simple diffusion does not involve a protein. Simple diffusion is classified as a means of passive transport, where a hydrophobic molecule (a nonpolar molecule that repels water) can penetrate the hydrophobic core of the phospholipid bilayer (a thin polar membrane made of two layers of lipid molecules) without getting rejected. Molecules that are substantially hydrophobic in nature are carbon dioxide, oxygen, and ethanol. An example of simple diffusion is osmosis.

. . . .

A substance (usually in lesser amount) dissolved in another substance, which is a solvent. A measure of how much of a solute is dissolved in a solvent is the concentration of a solute in a solution. The maximum quantity of solute that can dissolve in a specific volume of solvent varies with temperature. In a solution of sugar dissolved in water, sugar is the solute and water is the solvent.

. . . .

SOLVENT

· ·

A liquid, solid, or gas in which a solute (a chemically different liquid, solid, or gas) is dissolved, resulting in a solution. The solvent does the dissolving. The solution essentially takes on the characteristics of the solvent, including its phase, and the solvent becomes the major portion of the mixture. This is seen in solutions of fluids, where the solvent is present in greater amount than the solute. The most widely used solvent is water, and because it dissolves many substances it is regarded as the universal solvent. Organic solvents are typically used in detergents (citrus terpenes); in perfumes (ethanol); as paint thinners (e.g., toluene, turpentine); and as nail polish removers (acetone, methyl acetate, ethyl acetate). Aside from water, the use of inorganic solvents is fairly limited to research chemistry and certain technological processes.

SOMATIC CELL

· ·

SORUS

· ·

Any biological cell forming the body of an organism. The word *somatic* is derived from the Greek, meaning "body." In a multicellular organism, somatic cells are any cells apart from gametes (sperm and egg cells), germ cells (cells that give rise to gametes), and undifferentiated stem cells (cells that divide through mitosis and differentiate into diverse specialized cell types). Examples of somatic cells are cells of internal organs, skin, bones, blood, and connective tissues. Somatic cells are diploid, meaning they contain chromosomes arranged in pairs.

. . . .

A brownish or yellowish cluster of sporangia (structures producing and containing spores), usually located on the underside of fern leaves. In some species, sori (plural of sorus) are protected during development by an indusium, a scale or flap of tissue that forms an umbrella-like cover. As the sporangia mature, the indusium shrivels, leaving spore release unobstructed. The sporangia then burst and release the spores. To identify the fern taxa, or taxonomy, the shape, arrangement, and location of the sori can serve as valuable clues. Sori may be circular or linear, they may be arranged in rows or randomly, and their location may be marginal or set away from the border of the leaf.

. . . .

SPECIATION

The evolutionary process, or lineage-splitting event, in which two or more separate species are produced. Biologist Orator F. Cook first coined the term speciation for the "splitting of lineages" or "cladogenesis." One cause of speciation is geographic isolation; what was once a continuous population is divided into two or more smaller populations. Although this generally indicates a physical barrier separation, isolation could also occur from an unfavorable habitat between two populations that keeps them from mating with one another. Another cause of speciation is reduction of gene flow, a situation in which mating throughout a population is not random in a population that extends over a broad geographic range. Individuals on one end of the range would have zero chance of mating with individuals on the other end of the range.

. . . .

SPHYGMOMANOMETER

A device used to measure blood pressure. A sphygmomanometer is composed of an inflatable cuff to restrict blood flow, an instrument for inflation (a bulb and valve operated manually or a pump operated electrically), and a manometer (mercury or mechanical) to measure the pressure. The unit of measurement is millimeters of mercury (mmHg). A sphygmomanometer is also known as a *blood pressure meter* and a *sphygmometer*. There are two types of sphygmomanometers:

1. Manual sphygmomanometers are used in conjunction with a stethoscope for auscultation (listening to the internal sounds of the body). Used by trained practitioners, these sphygmomanometers cannot be used in a noisy environment, as it would distract from hearing distinctive sounds. There are two types of manual sphygmomanometers, mercury and aneroid.

2. Rather than auscultation, digital sphygmomanometers contain an electronic pressure sensor (transducer) to observe cuff pressure oscillations and automatically interpret them. They may use manual or automatic inflation and deflation of the cuff. As no stethoscope is needed, this device can be operated without training and can be used in noisy environments.

SPINDLE FIBER

....

SPIRACLE

....

One of the aggregates of microtubules (straight, hollow cylinders in the cytoskeleton) chiefly involved in moving and segregating chromosomes during cell division. During prophase of mitosis, these microtubules form at opposite poles of the cell and join together, forming a spindle-shaped structure. During metaphase of mitosis, the spindle fibers extend from the cell poles and attach to chromosomes. These spindle fibers then start to pull the chromosomes to opposite poles, marking the anaphase of cell division.

. . . .

An external respiratory opening. These openings allow the air to enter into the respiratory system of the body. A spiracle can appear as one of the openings in the exoskeleton of an insect, or as a pair of vestigial gill slits behind the eye of a cartilaginous fish (jawed fish with paired fins, scales, a two-chambered heart, and skeletons made of cartilage).

. . . .

SPLICING

A two-step biochemical process in which introns are removed from a primary RNA transcript and exons are joined together to form a mature messenger RNA (mRNA). An intron is a nucleotide sequence, or a portion of a precursor RNA. An exon is a nucleic acid sequence that is represented in the final mature RNA after RNA splicing. For many eukaryotic introns, splicing is done in a series of reactions that are catalyzed by the spliceosome, an RNA-protein complex composed of five small nuclear ribonucleoproteins (snRNPs). Self-splicing introns exist, but are rare. Eukaryotes mainly splice protein-coding mRNAs and some noncoding RNAs. Prokaryotes, which lack the spliceosomal pathway, splice rarely; however, when they do, they splice primarily noncoding RNAs. Splicing needs to occur before the mRNA can be used to produce a correct protein through translation.

. . . .

SPONGY MESOPHYLL

. . . .

SPONTANEOUS GENERATION

. . . .

A type of leaf tissue consisting of loosely arranged, chloroplast-bearing, lobed cells. In
the leaf, this tissue is part of the mesophyll,
where it forms a layer next to the palisade
cells. There are many intercellular air spaces
between the spongy mesophyll cells, which
allow for the interchange of gases (CO_2) that
are needed for photosynthesis and respiration. Spongy mesophyll cells contain less
chloroplast than palisade mesophyll cells, and
therefore are less likely to go through photosynthesis. Spongy mesophyll is also known as
spongy parenchyma.

. . . .

The obsolete theory that complex, living
organisms develop from nonliving matter. A
popular belief used to explain the origin of
life, spontaneous generation referred to the
supposed process by which life would system-
atically emerge from sources other than seeds,
eggs, or parents. Many believed in the theory
because it explained such occurrences as the
appearance of maggots on decaying meat. Ultimately, the ideas of spontaneous generation
were displaced in the 18th and 19th centuries
by the germ theory and the cell theory.

. . . .

SPORANGIOPHORE

. . . .

SPORANGIUM

. . . .

One of the asexual fertile components of the strobilus, a structure present on many land plant species. The sporangiophore consists of a transparent stalk bearing a flattened disk at its apex, on the lower edge of which is a ring of 5 to 10 fingerlike sporangia. Each sporangium opens and sheds large numbers of thin-walled, green spores by a longitudinal slit on its inner side.

. . . .

A single-cell or multicellular enclosure in which spores are produced. Sporangia are associated with leaves or they can be terminal (on the tips of stems) or lateral (placed along the sides of stems). Sporangia produce haploid spores by meiosis. All plants and fungi form sporangia at some point in their life cycles.

. . . .

SPOROPHYTE

· ·

STATOCYTE

· ·

An asexual, spore-bearing generation in plants. In many plants, meiosis and fertilization divide the life of an organism into two parts, or generations. In a cycle known as alternation of generations, the haploid, gamete-producing generation is called gametophyte and the diploid, spore-producing generation is called sporophyte. Sporophytes are the large, nutritionally independent, and complex generation in vascular plants. A more dominant generation than the gametophyte, the sporophyte serves as the storage unit of the cell's genetic information. The gametophyte begins with a spore, produced as a result of meiosis. By mitosis, the spore produces gametes, and via sexual reproduction, later produces the sporophyte. The sporophyte produces haploid spores as a result of meiosis. These spores then undergo mitotic division, giving rise to gametophytes, which later produce gametes. The union of these gametes forms diploid zygotes. The zygotes then divide mitotically to form new sporophytes.

. . . .

A gravity-perceiving cell located in the root cap, a section of tissue at the tip of a plant root. Statocytes contain starch-filled nonpigmented organelles called statoliths, which sediment at the lowest part of the cells and initiate differential growth patterns, bending the root toward the vertical axis.

. . . .

STELE

. . . .

STIPE

. . . .

The central core of the stem and root of a vascular plant, containing the vascular tissues derived from the procambium (xylem and phloem).

. . . .

A stalk that supports some other. The precise meaning of a stipe varies depending on which taxonomic group is being described. In flowering plants, a stipe is a stalk that occasionally supports a flower's ovary. In the case of ferns, the stipe is only the stalk attaching the stem to the beginning of the leaf tissue. A stipe is also found in other organisms that are not plants, like mushrooms and seaweed, and is particularly common among brown algae such as kelp. The stipe of a kelp contains cells that serve to transport nutrients within the alga, similar to the function of phloem in vascular plants.

. . . .

STOLON

. . . .

A specialized type of horizontal above-ground shoot, a colonizing organ that arises from an axillary bud near the base of the plant. A stolon has thin internodes and a propensity to form adventitious roots at the nodes, which are roots that arise out of sequence from the more usual root formation of branches of a primary root. Plants with stolons are called stoloniferous. The most common example of a stolon is the strawberry (genus *Fragaria*), in which the mother plant forms plantlets on stolons during growth. In the case of the strawberry, the stolon is called a runner, which is not a self-sustaining structure, but only a connector between ramet (an individual) and mother plant.

. . . .

STOMA

. . . .

S

A tiny pore found in the leaf and stem epidermis that serves as the site for gas exchange. The stoma is bordered by a pair of guard cells that regulate its opening and closure. The term *stomata*, Greek for "mouth," is used to refer to both the pore itself and its accompanying guard cells. Air containing carbon dioxide and oxygen enters and exits the plant through these openings, where it is used in photosynthesis and respiration, respectively. In transpiration, water vapor is also released into the atmosphere through these pores.

. . . .

STRATIFIED EPITHELIUM

Multiple layers of epithelial cells cemented together over the external surface of the body and lining most of the hollow structures. Stratified epithelia cover the exterior body surfaces and line portions of the body tracts where friction phenomena occur. There are various subtypes of stratified epithelium, named for the type of cells on the surface:

- Stratified squamous epithelia are composed of squamous (flat), scalelike, multilayered epithelial cells over a basement membrane (thin sheet of fibers that line the cavities and surfaces of organs). As the thickest type of epithelium, it is well suited to areas subject to constant abrasion, and layers can be sloughed off and replaced before the basement membrane is exposed. The stratified squamous epithelia form the outermost layer of the skin and the inner lining of the mouth, esophagus, and vagina. It is further classified by the presence of keratin, a tough protective protein, in the skin, tongue, and external portion of the lips.

- Stratified columnar epithelium is a rare type of epithelial tissue composed of multilayered, column-shaped cells that function in secretion and protection. Stratified columnar epithelia are found in the ocular conjunctiva of the eye, near the salivary glands, and in parts of the pharynx and anus, the uterus, the urethra, and the vas deferens.

- Stratified cuboidal epithelium is also a rare type of epithelial tissue; it is composed of multilayered, cube-shaped cells that protect areas of larger ducts such as sweat glands in the skin, mammary glands in the breast, and in the mouth.

STROBILIS

....

STROMA

....

SUBSTRATE

A reproductive structure featured on many land plant species that consist of sporangia-bearing structures densely aggregated along a stem. Strobili are characterized by a central axis (a stem) surrounded by modified leaves (leaves adapted to perform functions other than photosynthesis) or modified stems (stems that can be found either above or below ground, such as bulbs and runners). Leaves that bear sporangia are called *sporophylls*, whereas sporangia-bearing stems are called *sporangiophores*. The cones of pine trees are considered strobili.

. . . .

The supporting framework or structural tissue of an organ, usually composed of connective tissue cells. Stromata support the function of the parenchymal cells of an organ or body part (the main functional cells of an organ). The most common types of stromal cells are fibroblasts, immune cells, and inflammatory cells.

. . . .

The surface on which a plant or animal grows or where it is attached; the earthy material in which an organism lives. A substrate can include biotic (living) or abiotic (nonliving) materials and animals. For example, the rock that an alga lives on is considered its substrate. Furthermore, the alga can itself serve as a substrate for another animal that lives on top of it.

SULFIDE

....

SUPERNATANT

....

Any of three classes of chemical compounds containing the element sulfur. One class is phosphine sulfides, formed when a sulfur atom is linked to phosphorus by a bond that has both ionic and covalent properties. Another class is inorganic sulfides, which are ionic compounds containing the negatively charged sulfide ion, S^{2-}. Compounds in which a sulfur atom is covalently bonded to two organic groups comprise the third class of organic sulfides. Organic sulfides are sometimes called thioethers, and are well known for their bad odors. Many important naturally occurring metal ores are sulfides. Significant examples include zinc, cadmium, mercury, copper, silver, and pyrite (fool's gold).

· ·

The liquid fraction deposited above a solid residue after crystallization, precipitation, centrifugation, or other process. Supernatant is also known as *supernate*.

· ·

SURFACTANT

....

SYMBIONT

....

A substance that can greatly reduce the surface tension of a liquid, the interfacial tension between two liquids, or that between a liquid and a solid. At the surface of water, molecules hold on to each other tightly because there are no molecules pulling on them from the air above. As the molecules on the surface stick together, they form an invisible skin called surface tension. Interfacial tension occurs between the liquid phase of one substance and either a solid, liquid, or gas phase of another substance. The interaction occurs at the surfaces (interfaces) of the substances involved. A major component of cleaning products, surfactants may act as detergents, emulsifiers, wetting agents, and foaming agents. A surfactant lowers the surface tension of the medium in which it is dissolved. For example, the lower surface tension of water makes it easier to lift dirt off dirty dishes, and helps to keep the particles suspended in the dirty water. Each surfactant molecule is made up of a hydrophilic (water-soluble) head that is attracted to water molecules and a hydrophobic (water-insoluble) tail that repels water and simultaneously attaches itself to oil and grease in dirt. These opposing forces release the dirt and suspend it in the water.

. . . .

An organism that lives in a mutually beneficial relationship or in symbiosis with another organism.

. . . .

SYMBIOSIS

S

A close and often long-term interaction between the individuals of two or more different species. Ecologists use a different term for each type of symbiotic relationship:

- Mutualism is the relationship where both individuals derive a benefit. An example is bacteria that live within the intestines of certain mammals. The mammals benefit from the enzymes that the bacteria produce to facilitate digestion, and the bacteria benefit from having a stable supply of nutrients in the host environment.

- Commensalism is the relationship where one individual benefits and the other is not significantly harmed or helped. An example is when barnacles attach themselves to the shell of a scallop; the attachment leaves the scallop unaffected, but the barnacles benefit by having a place to stay.

- Parasitism is the relationship where one organism (the parasite) benefits at the expense of another organism (the host). The close association may lead to the injury of the host. An example is a human and a tapeworm living in the person's intestines. The tapeworm derives food and shelter from the human host while the human is denied the nutrition that is consumed by the tapeworm.

SYMPATRIC SPECIATION

. . . .

The origin of new species in populations that overlap geographically. Sympatric speciation is the process through which new descendant species evolve from a single ancestral species while all are inhabiting the same geographic region. It is one of three geographic categories of speciation. Sympatric speciation is common in plants. For instance, parent plants produce polyploid offspring (containing more than two paired sets of chromosomes) that live in the same environment as their parents but are reproductively isolated. However, sympatric speciation is still a highly contentious issue among biologists, who question whether this type of speciation happens very often despite existing empirical evidence.

. . . .

SYNAPSE

S

A structure in the nervous system that allows a neuron to pass an electrical or chemical signal to another cell (neural or otherwise). Synapses are essential to neuronal function. There are two types of synapses:

1. Chemical synapse: The presynaptic neuron releases a neurotransmitter (a chemical) that binds to receptors (molecules that allow cells to communicate with one another through chemical signals). The receptors are embedded in the plasma membrane of the postsynaptic cell. The neurotransmitter, which transmits signals from a neuron to a target cell across a synapse, may either excite or inhibit the postsynaptic neuron.

2. Electrical synapse: Presynaptic and postsynaptic cell membranes are connected by gap junctions (channels) that pass electric current from the presynaptic cell to the postsynaptic cell. This creates a rapid transfer of signals from one cell to the next.

SYNTHETASE

. . . .

SYSTEMIC CIRCULATION

S

An enzyme that catalyzes the joining together of two large molecules by using the energy derived from the breakdown of a pyrophosphate bond in ATP or a similar triphosphate. Synthetase is used to synthesize new molecules. Synthetase is also known as *ligase*.

. . . .

The part of the cardiovascular system in which oxygenated blood leaves the heart, supplies nourishment to all of the tissue in the body, and returns deoxygenated blood back to the heart. Arteries, veins, and capillaries are the blood vessels responsible for the delivery of oxygen and nutrients to the tissue. In systemic circulation, oxygen-rich blood leaves through the left ventricle to the aorta, the heart's main artery, which branches into smaller blood vessels (arterioles and capillaries) that run throughout the body. The oxygen-rich blood enters the capillaries where the oxygen and nutrients are released. The waste products are collected and the waste-rich blood flows into the venous capillaries and then into the venae cavae, through which the blood reenters the heart at the right atrium. In a phase of systemic circulation called renal circulation, blood passes through the kidneys, where much of the waste is filtered from the blood. Blood also passes through the small intestine in a phase called portal circulation, where the blood from the small intestine collects in the portal vein and passes through the liver. The liver then filters sugars from the blood, storing them for later.

TARGET CELL

....

TATA BOX

....

Abnormal red blood cells that have the appearance of a shooting target with a bull's-eye. Target cells have a dark center (a hemoglobinized area) enclosed by an area of pallor (a white ring), followed by a dark outer second ring containing a band of hemoglobin. These cells are associated with liver disease, hemoglobin C disease, and severe iron deficiency anemia. Target cells are also known as *codocytes*.

. . . .

A DNA sequence, usually TATAAA, which indicates where a genetic sequence can be read and decoded. Found in the promoter region of genes in archaea and eukaryotes, the TATA box is considered to be the core promoter sequence, specifying to other molecules where transcription begins, defining the direction of transcription, and indicating the DNA strand to be read. Many eukaryotic genes have a conserved TATA box located 25 to 35 base pairs before the transcription start site of a gene. The TATA box can bind to proteins called transcription factors and can recruit the enzyme RNA polymerase, which synthesizes RNA from DNA. The TATA box is also known as the *Goldberg-Hogness box*.

. . . .

TAXONOMY

....

T CELL

....

T

The conception, naming, and classification of organism groups. The groups created through taxonomy are referred to as taxa (singular: taxon). Each taxon is designated a taxonomic rank and placed in a systematic hierarchy reflecting evolutionary relationships. The eight major taxonomic ranks are species, genus, family, order, class, phylum (or division), kingdom, and domain. There also are intermediate minor rankings between the main ones, which can be produced by adding prefixes such as *super-*, *sub-*, or *infra-*. For example, a subclass has a rank between class and order, and a superfamily has a rank between order and family.

. . . .

One of the cells that belong to a group of white blood cells called lymphocytes and that play a central role in adaptive immunity, the system that tailors the body's immune response to specific pathogens. After they are produced in the bone marrow, T cells mature and develop in an organ in the chest called the thymus (what T cells are named after). After maturation, T cells are present in the blood and in lymph nodes. White blood cells protect the body from infection, and the function of T cells is to fight infection. T cells may be affected in cancers like lymphoma and in diseases of the immune system like HIV/AIDS. T cells are also known as *T lymphocytes*.

. . . .

TELOMERASE

. . . .

T

An enzyme made of protein and RNA subunits that add a repetitive DNA sequence (TTAGGG) to caps of DNA located at the ends of existing chromosomes called telomeres. As a reverse transcriptase (a DNA polymerase enzyme that transcribes single-stranded RNA into single-stranded DNA), telomerase carries its own RNA molecule, which is used as a template when it lengthens telomeres. Telomerase is found in adult germ cells, tumor cells, and fetal tissues. Telomerase activity is regulated during development and has very low activity in body cells, which makes them age, resulting in an aging body. If telomerase is activated in a cell, the cell will grow and divide virtually forever. Consequently, the enzyme has recently been found in many human tumors and it has been suggested that telomerase therapies may be used to combat cancer and to extend life span significantly.

· ·

TELOMERE

· ·

TELOPHASE

· ·

A region of repetitive DNA located at the end of a chromosome. Telomeres function as disposable buffers, protecting the end of chromosomes from deterioration or from fusion with neighboring chromosomes. A telomere can reach a length of 15,000 base pairs. However, due to cell division, some of the telomere can be lost (25 to 200 base pairs per division). When the telomere becomes too short, the chromosome can no longer replicate, and a cell ages and eventually dies by a process called apoptosis. Telomere activity is controlled by two mechanisms: erosion, which occurs each time a cell divides, and addition, which is determined by the activity of telomerase.

. . . .

The fifth and final phase of mitosis. Telophase begins after anaphase, when sister chromatids (two identical copies of a chromatid connected by a centromere) separate from one another, becoming separate daughter chromosomes that are pulled to opposite poles of a cell. During telophase, two new nuclear membranes begin to form around each of the two separated sets of chromosomes, separating the nuclear DNA from the cytoplasm. The chromosomes begin to uncoil, making them decondense and become more diffuse.

. . . .

AP* BIOLOGY FLASH REVIEW

TERMINAL BUD

. . . .

THERMAL INVERSION

. . . .

The bud of a plant located at the apex (tip) of a stem. The terminal bud marks the end of that year's growth and the starting point of the following year's growth. It is different from a lateral bud, which grows from the side. The terminal bud is the main area of growth in most plants. It is the dominant bud, since it can inhibit the growth of axillary (lateral) buds. Terminal buds have specialized tissue called apical meristem, which are composed of cells that can divide indefinitely. These cells produce all differentiated tissue, including reproductive and vegetative organs. Surrounding the terminal bud is a complex arrangement of nodes and internodes with maturing leaves. The terminal bud is also known as the *apical bud*.

Condition in which a layer of warm air overlies a layer of cooler air that settles near the ground. The warm layer acts as a cap, trapping the surface air in place and preventing dispersion of any pollutants it contains. As a result, pollution such as smog can be trapped near the ground. Thermal inversion is a deviation from the normal change of an atmospheric property with altitude; usually air becomes cooler as altitude increases. Thermal inversion blocks atmospheric flow, causing the air over an inversion area to become stable. Unblocking can result in weather patterns such as freezing rain and intense thunderstorms and tornadoes. Thermal inversion is also known as *temperature inversion*.

THIGMOTROPISM

································

THRESHOLD POTENTIAL

································

T

A plant that moves or grows in response to touch or physical contact with a solid object. The prefix *thigmo* comes from the Greek for "touch." Thigmotropism is illustrated by the climbing tendrils of some plants. The tendrils sense the solid object, grow toward the touch stimulus, and develop tendrils that wrap or coil around the object. Roots also depend on touch sensitivity to grow through soil. However, when a root senses an object, it navigates away from the object, and is said to be negatively thigmotropic. This allows the roots to go through the soil with minimum resistance.

. . . .

The critical level of depolarization of the plasma membrane to be reached for initiation of action potential. Threshold potential is the number voltage difference has to be in order to initiate an action potential, which is when the difference in electrical potential between the interior and the exterior of a biological cell rapidly rises and falls. The threshold potential is typically between -40 and -55 mV.

. . . .

THYLAKOID

. . . .

TITRATION

. . . .

One of the membrane-bound compartments inside chloroplasts and cyanobacteria, a phylum of bacteria that obtain their energy through photosynthesis. The light-dependent reactions of photosynthesis take place in the thylakoids and are concerned with the initial conversion of light energy into chemical energy stored in ATP and NADPH.

. . . .

An operation, used in volumetric analysis, to determine the concentration of a substance in solution. Titration is a process in which a measured amount of one solution is added to a standard reagent of known concentration until a reaction of definite and known proportion between the two is complete. If the concentration of one solution is known, the concentration of the other can be determined. A reagent (titrant) is prepared as a standard solution. A known concentration and volume of the reagent reacts with a solution of the substance undergoing analysis (titrand) to determine concentration. Titration is also known as *titrimetry*.

. . . .

T-MAZE

. . . .

TONOPLAST

. . . .

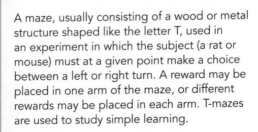

A maze, usually consisting of a wood or metal structure shaped like the letter T, used in an experiment in which the subject (a rat or mouse) must at a given point make a choice between a left or right turn. A reward may be placed in one arm of the maze, or different rewards may be placed in each arm. T-mazes are used to study simple learning.

· ·

The cytoplasmic membrane surrounding a vacuole that separates the vacuolar contents from the the cytoplasm in a cell. As a membrane, the primary function of the tonoplast is to regulate the movements of ions around the cell and isolate materials that might be harmful or pose as a threat to the cell. The tonoplast is also known as the *vacuolar membrane*.

· ·

TRACHEID

. . . .

T

A single elongated cell in the xylem (fluid-conducting tissue) of vascular plants that serves as structural support but primarily functions in the transport of water and mineral salts collected by the roots to other parts of the plant (e.g., stem, leaves, fruits). Tracheids are one of two types of tracheary elements of vascular plants, the other being vessel elements. At maturity, the protoplast of a tracheid cell has broken down and disappeared. The cell then develops a secondary cell wall composed of cellulose thickened with lignin (a chemical binding substance). Tracheids have pits on their end walls that interconnect them with neighboring tracheid cells so that water can move across from one cell to another.

. . . .

TRANSCRIPTION

T

A process of creating a complementary RNA copy from a sequence of DNA (a template). Whether prokaryotic or eukaryotic, transcription, has three main steps:

1. Initiation is the binding of an enzyme called RNA polymerase to double-stranded DNA. Specific nucleotide sequences tell RNA polymerase where to begin and end. RNA polymerase binds to a sequence of DNA at an area called the promoter region.

2. During elongation, transcription factors (other proteins required for transcription) unwind the DNA strand and allow RNA polymerase to transcribe only a single strand of DNA into a single-stranded RNA polymer (a large molecule composed of repeating structural units) called messenger RNA (mRNA).

3. During termination, RNA polymerase travels along the DNA until it reaches a transcription terminator, a section of genetic sequence that marks the end of transcription. RNA polymerase then releases the mRNA polymer and detaches from the DNA.

. . . .

TRANSDUCTION

．．．．．．．．．．．．．．．．．．．．．．．．．．．．．．．．

TRANSITION STATE

．．．．．．．．．．．．．．．．．．．．．．．．．．．．．．．．

The process by which DNA is transferred from the genes of one bacterium to another by a virus. Transduction, which does not require cell-to-cell contact, is a tool used by molecular biologists to introduce a foreign gene into a host cell's genome (an organism's hereditary information, including its genes and noncoding sequences of DNA or RNA). This can be done via a viral vector, such as a retrovirus (an RNA virus that is duplicated in a host cell using the reverse transcriptase enzyme that transcribes single-stranded RNA into single-stranded DNA) or a bacteriophase (virus that injects its viral DNA into bacteria).

· ·

During a chemical reaction, the highest energy formed during the transition from reactants to products. During this transition, the energy of the system increases. As the reaction encounter continues, the potential energy goes on increasing until the system reaches a structure of maximum energy. This highest point on the reaction coordinate (an abstract one-dimensional coordinate that represents progress along a reaction pathway) is the transition state.

· ·

TRANSLOCATION

A rearrangement of parts between nonhomologous chromosomes as a result of abnormal breakage and refusion of reciprocal segments. There are two main types of translocation. Robertsonian translocation, usually involving two acrocentric chromosomes, is when breakage occurs at the centromeres and entire chromosome arms are exchanged. Reciprocal translocation (also known as non-Robertsonian) is a mutual exchange of parts between two broken nonhomologous chromosomes, where one part of one chromosome unites with one part of the other chromosome. Translocations can be balanced (an even exchange of material without loss or gain) or unbalanced (an unequal exchange of material resulting in extra or missing genes). Conditions such as cancer, infertility, and Down syndrome can be caused by translocation.

TRANSPIRATION

T

Part of the water cycle, it is the evaporation of water into the atmosphere from the leaves and stems of plants. Transpiration accounts for approximately 10% of all evaporating water. For transpiration to occur, plants must first absorb soil water through their roots and pump the water up from the soil to deliver nutrients to their leaves. This pumping is driven by the evaporation of water through small pores or stoma, located on the undersides of leaves, and bordered by guard cells (together known as stomata) that open and close the pores. Transpiration occurs at the leaves while their stomata are open for the passage of carbon dioxide and oxygen during photosynthesis. Environmental factors that affect the rate of transpiration include light, temperature, humidity, wind, soil water, and number of leaves and stomata.

. . . .

TRANSPORT EPITHELIUM

. . . .

One or more layers of specialized epithelial cells that regulate solute movements. Transport epithelia allow salts and other wastes to move from blood vessels to secretory tubules. Osmotic regulation and metabolic waste disposal in most animals depend on the capacity of transport epithelia to move specific solutes in controlled amounts in specific directions. The types and directions of solutes that move across the transport epithelium are determined by the molecular structure of plasma membranes. Transport epithelia are arranged into complex tubular networks with extensive surface area.

TRANSPOSABLE ELEMENT

• •

A DNA sequence that can change its relative position (self-transpose); it can move itself or a copy of itself from one place to another within a cell's genome. Because transposable elements can move to different chromosomes, they contribute to the creation of new genes. Often considered junk DNA, transposable elements are portions of the DNA sequence for which no function has been identified. Transposable elements are assigned to one of two classes according to their mechanism of transposition: class I (retrotransposons), which is a form of copy-and-paste transposition mechanism; and class II (DNA transposons), which is a form of cut-and-paste transposition mechanism. Transposable elements are also known as *transposons* or *jumping genes*.

. . . .

TRICHOME

One of the fine outgrowths or appendages that are of diverse structure and function in plants. Trichomes are produced from the external growth of epidermal cells on leaf or stem surfaces. These appendages give plant leaves or stems distinctive textures such as spines, hairs, or glands. Trichomes may be unicellular or multicellular, branched or unbranched. Branched trichomes can be dendritic (treelike), tufted, or stellate (star-shaped). As root hairs, trichomes absorb water and minerals. As leaf hairs, they provide defense against insects and small herbivores, lower plant temperature, reflect radiation, and reduce water loss.

· ·

TRIGLYCERIDE

The major form of fat stored by the body. Binding one molecule of the alcohol glycerol with three molecules of fatty acid produces a triglyceride. Triglycerides are not only produced in the body but also come from the foods we eat. They are a mechanism for storing unused calories, and their high concentration in blood is associated with the intake of starchy carbohydrate foods. The molecular composition of a triglyceride can vary. Saturated triglycerides contain fatty acids that have no double bonds, only single carbon bonds (C–C–C). Elevated levels of saturated triglycerides can raise blood cholesterol, which, in turn, can lead to atherosclerosis (hardening of the arteries), heart disease, stroke, and pancreatitis (inflammation of the pancreas). Animal fats found in meat, poultry, and whole milk dairy products are high in saturated fats. Unsaturated triglycerides contain fatty acids that have one double bond (C=C). Foods high in unsaturated fat are avocados and a variety of oils (olive, peanut, sesame seed, and canola). Polyunsaturated triglycerides contain fatty acids that have multiple double bonds (C=C=C). Examples of polyunsaturated fats are corn oils, sunflower oils, soybean oils, and mayonnaise.

. . . .

TRISOMY

. . . .

A condition where the presence of a single extra chromosome yields a total of three copies instead of the normal two. A trisomy is a type of chromosome abnormality or aneuploidy (an abnormal number of chromosomes). During cell division, if chromosome pairs do not separate correctly, the egg or sperm may end up with a second copy of one of the chromosomes. If fertilization occurs, the resulting embryo may also contain a copy of the extra chromosome. Full trisomy occurs when an entire extra chromosome has been copied. Partial trisomy refers to an extra copy of a segment of a chromosome. Autosomal trisomies, which are trisomies of the nonsex chromosomes, are referenced by the specific chromosome that has an extra copy. For example, the presence of an extra chromosome 21 is called trisomy 21, which is found in Down syndrome.

. . . .

TROPHIC LEVEL

. . . .

The feeding position that an organism occupies in a food chain. The word *trophic* derives from the Greek for "food" or "feeding." A food chain represents a sequence of organisms in an ecological community that transfer food energy from one organism to another as each consumes a lower member on the chain and in turn is preyed upon by a higher member on the chain. The number of steps an organism is from the start of the chain is a measure of its trophic level. Trophic levels can be represented by numbers:

- Level 1: Plants and algae are primary producers (organisms that can make their own food).
- Level 2: Herbivores that are primary consumers that eat primary producers, like plants. Consumers are animals that cannot manufacture their own food and need to consume other organisms.
- Level 3: Carnivores that are secondary consumers eat other animals that are herbivores.
- Level 4: Carnivores that are tertiary consumers eat other carnivores.
- Level 5: Apex predators, which have no predators of their own, are at the top of the food chain.

TROPOMYOSIN

A long protein strand that binds along the length of the actin filament and regulates actin mechanics. Tropomyosin, together with troponin complex (a complex of three regulatory proteins), controls the interaction of actin and myosin, which are proteins important in muscle contraction. In the absence of a nerve impulse, the tropomyosin lies within the groove between actin filaments in muscle tissue, blocking the myosin-binding sites in actin. At this point, the muscle is relaxed or at rest. In the presence of nerve impulse, calcium channels open in the sarcoplasmic reticulum, causing the release of calcium ions. The calcium then binds to troponin, causing a conformational change (a change in the shape of a protein as a result of a change in the environment) that moves tropomyosin out of the way. This allows myosin to bind with actin molecules, resulting in muscle contraction.

. . . .

TRYPANOSOME

Any of a group of unicellular parasitic flagellate protozoa of the genus *Trypanosoma*, which are parasites of insects, plants, birds, bats, fish, amphibians, and mammals. All trypanosomes are heteroxenous (requiring more than one host to complete life cycle) and are transmitted via a vector, usually by the bite of blood-feeding insects. Trypanosomes, which have existed for over 300 million years, can cause serious diseases such as sleeping sickness, which results in swelling of the brain. Two different types of trypanosomes exist:

1. Stercorarian trypanosomes develop in the posterior gut of an insect, usually the triatomid kissing bug, and are excreted in the feces of the insect and onto the skin of the host. They then penetrate the skin and disseminate throughout the body.

2. Salivarian trypanosomes develop in the anterior gut of their vector, the tsetse fly, and are then inoculated into a mammalian host through a bite before a blood meal.

. . . .

TUBULIN

. . . .

TURGOR PRESSURE

. . . .

A globular protein that is the basic structural constituent of microtubules of living cells. The most common tubulins are the two proteins that make up microtubules: α-tubulin (alpha) and β-tubulin (beta). Each has a diameter of 4 to 5 nanometers and a molecular mass of approximately 55 kilo daltons, the standard unit that is used for indicating mass on an atomic or molecular scale.

. . . .

The main pressure exerted by water inside the cell against the cell wall in plant cells. Turgor pressure is caused by the osmotic flow of water from an area outside of the cell with low solute concentration into the cell's vacuole, which has a higher solute concentration. This influx of water results in turgor pressure, which pushes the plasma membrane against the cell wall. This force gives the plant rigidity, helping it to stand upright. Turgor can result in the bursting of a cell. Turgor pressure is also known as *turgidity*.

. . . .

UNCOUPLER

. . . .

VACUOLE

. . . .

An agent capable of dissociating two integrated series of chemical reactions. In particular, an uncoupler is an agent that can prevent ATP synthesis by uncoupling electron transport from oxidative phosphorylation (a metabolic pathway that uses energy released by the oxidation of nutrients to produce ATP).

. . . .

A membrane-bound organelle found in the cytoplasm of all plant and fungal cells and of some protist, animal, and bacterial cells. Filled with water, vacuoles are enclosed sacs that contain inorganic and organic molecules. A vacuole's structure varies according to the needs of the cell; therefore, it has no standard shape or size. Vacuoles are small in animal cells, but large in plant cells. The functions of the vacuole include intracellular storage, secretion, digestion, and excretion.

. . . .

VENULE

. . . .

Any of the very small blood vessels connecting capillaries with larger systemic veins. Venules range in size from 8 to 100 μm (micrometers) in diameter. They are formed when capillaries unite; when venules unite, they form a vein. Venules serve as a passageway for deoxygenated blood returning from the capillaries to the vein. They have a three-layer wall: the innermost layer is comprised of squamous endothelial cells that act as a membrane; the middle layer contains muscle and elastic tissue; and the outermost layer consists of fibrous connective tissue. Fluids and blood cells can move easily through venule walls because they are thin and very porous.

. . . .

VILLUS

....

The many tiny, fingerlike structures that stick out from the epithelial lining of the intestinal wall. Villi (plural of villus) greatly increase the internal surface area of the intestinal walls, which allow for increased intestinal wall area available for food absorption. Villi number about 6,000 to 25,000 per square inch of tissue, and each villus is approximately 0.5 to 1.6 millimeters in length. Villi move in swaying, contracting motions that increase the flow of blood and lymph to enhance absorption. Villi absorb about 2 gallons (7.5 liters) of fluid per day. Each villus has a central core composed of one artery and one vein, a strand of muscle, a centrally located lymphatic capillary (a vessel that absorbs dietary fats from the gut into the bloodstream), and connective tissue that adds support to the structures. A surface mucous-membrane layer covers the core of a villus, helping to secrete mucus into the intestinal cavity and to absorb the substances passed into the blood and lymphatic vessels.

. . . .

XYLEM

X

A type of transport tissue in vascular plants. Xylem, which is the primary component of wood, is derived from the Greek word meaning "wood." Within xylem are tracheids, elongated cells that serve in the transport of water and soluble mineral nutrients throughout the plant. Xylems also function to replace water lost during photosynthesis and transpiration. There are two forms of xylem:

- Primary xylem is formed during primary growth from procambium, a type of meristem (a tissue made of undifferentiated cells). A plant that does not form a woody stem (i.e., nonwoody plants, such as an herb) contains only the primary xylem. Primary xylem consists of small tracheids and vessels.

- Secondary xylem is formed during secondary growth from vascular cambium (a lateral meristem). These are tissues that cover the primary xylem, causing the primary xylem cells to die, lose their conducting function, and form a hard skeleton that serves only as support for the plant. Secondary xylem is present in trees and shrubs. Its cell walls are thickened by lignin (a chemical compound that adds strength and stiffness to the cell walls), which provides mechanical support. Secondary xylem consists of large tracheids and vessels.

ZONE OF CELL
DIFFERENTIATION

. . . .

Z

The region that is located at the top of the root, closest to the plant. The zone of cell differentiation is where cell differentiation takes place, a process that consists of a less specialized cell becoming a more specialized cell type. It is here that the cells of the root form the assortment of tissues that will make up the mature root. In this region, the xylem cells are the first of the vascular tissues to differentiate. Root hairs, which are used to absorb water and nutrients, are also found in this region. The zone of cell differentiation is also known as the *zone of maturation*.

. . . .

ZONE OF CELL DIVISION

. . . .

The area just above the apical meristem. The zone of cell division is where mitosis occurs, producing new cells. This region includes the apical meristem, which is undifferentiated tissue that begins growth of new cells at the tips of roots and shoots. The apical meristem is located in the center of the root tip, in an area protected by the root cap (a mass of cells that cover and protect the growing cells at the end of a root). The apical meristem of the root forms three primary meristems: protoderm, which produces cells that will become dermal tissue (epidermis); procambium, which produces cells that will become vascular tissue (xylem and phloem); and the ground meristem, which produces cells that will become ground tissue (cortex). This zone also contains a region called the quiescent center, where cells are not mitotically active and rarely divide. These cells, which are resistant to harmful factors, serve as a supply of healthy cells and can assist with the formation of a new apical meristem if one is damaged by external forces.

· ·

ZYGOMYCETE

. . . .

ZYGOSPORANGIUM

. . . .

A fungus that forms zygospores (a plant spore derived from the union of two similar sexual cells). Approximately 1,060 species of zygomycete are known. They live on decaying plant and animal matter in soil. Some are parasites of plants, insects, and small animals, while others form symbiotic relationships with plants. The life cycle of zygomycetes can be broken up into sexual and asexual reproductive phases.

. . . .

A thick-walled resistant spherical spore formed during the sexual reproduction of zygomycetes. Zygosporangia develop from the fusion of gametangia (an organ or a cell in which gametes are produced). The zygosporangial wall consists of a thin outer layer, which usually wears away, disappears, or is disrupted by a thicker, expanding inner wall layer, within which a zygospore develops. An example of a zygosporangium is the black bread mold, a common type of fungus.

. . . .

ZYGOTE

Z

AP* BIOLOGY FLASH REVIEW

[547]

A thick-walled, resistant spore resulting from the fusion of two similar gametes. A zygospore is a spore that is produced and contained within a zygosporangium. Young zygospores are irregular in shape and have thin walls. As its wall becomes thicker, the zygospore becomes more spherical in shape and remains closely associated with the wall of the parent zygosporangium. A zygospore remains dormant until the environment is favorable (light, moisture, heat, or chemicals secreted by plants); it then undergoes zygotic meiosis upon germinating, and haploid vegetative cells are released.

· ·

NOTES

NOTES

_____

_____

_____

_____

_____

_____

_____

_____

_____

_____

_____

_____

_____

_____

_____

_____

_____

NOTES

NOTES